OPEN SPACE: PEOPLE SPACE

Open Space: People Space offers a rare insight into people's engagement with the outdoor environment and looks at the ways in which design of spaces and places meets people's needs and desires in the twenty-first century. Embracing issues of social inclusion, recreation and environmental quality, it explores innovative ways to develop an understanding of how the landscape, urban or rural, can contribute to health and quality of life.

Open Space: People Space explores the nature and value of people's access to outdoor environments. Led by Edinburgh's OPENspace research centre and including contributions from international leaders in their fields, the debate focuses on current research to support good design for open space and brings expertise from a range of disciplines to look at:

- an analysis of policy and planning issues and challenges
- understanding the nature and experience of exclusion
- the development of evidence-based inclusive design
- the innovative research approaches which focus on people's access to open space and the implications of that experience.

Open Space: People Space is of value to policy makers, researchers, urban designers, landscape architects, planners, managers and students, and will also prove invaluable for those working in child development, health care and community development.

Catharine Ward Thompson is Research Professor of Landscape Architecture at Edinburgh College of Art. She is Director of OPENspace based at Edinburgh College of Art and Heriot-Watt University. Her award-winning research on historic urban parks and contemporary needs has led to projects exploring children's and young people's landscapes, the importance of outdoor access for older people and salutogenic environments for all.

Penny Travlou is Research Fellow at OPENspace and Lecturer in Cultural Geography and Visual Culture at the Centre for Visual and Cultural Studies at Edinburgh College of Art. Her work includes, among other themes, research on young people's perceptions and use of public open space in Edinburgh, supported by the Carnegie Trust and the British Academy, as well as broader explorations of young people's engagement with outdoor places.

OPEN SPACE: PEOPLE SPACE

Edited by Catharine Ward Thompson and Penny Travlou

Taylor & Francis
Taylor & Francis Group
LONDON AND NEW YORK

First published 2007
by Taylor & Francis
2 Park Square, Milton Park, Abingdon, Oxon, OX14 4RN

Simultaneously published in the USA and Canada
by Taylor and Francis Inc.
711 Third Ave, New York NY 10017

Taylor and Francis is an imprint of the Taylor & Francis Group, an informa business

Transferred to Digital Printing 2009

© 2007 Catharine Ward Thompson and Penny Travlou selection & editorial matter; individual contributions © the contributors

Typeset in Avenir by Keystroke, 28 High Street, Tettenhall, Wolverhampton

All rights reserved. No part of this book may be reprinted or reproduced or utilised in any form or by any electronic, mechanical, or other means, now known or hereafter invented, including photocopying and recording, or in any information storage or retrieval system, without permission in writing from the publishers.

British Library Cataloguing in Publication Data
A catalogue record for this book is available from the British Library

Library of Congress Cataloging in Publication Data
A catalog record for this book has been requested

ISBN10: 0–415–41533–0 (hbk)
ISBN10: 0–415–41534–9 (pbk)
ISBN10: 0–203–96182–X (ebk)

ISBN13: 978–0–415–41533–0 (hbk)
ISBN13: 978–0–415–41535–7 (pbk)
ISBN13: 978–0–203–96182–7 (ebk)

Contents

List of contributors   vii

Foreword
Laurie Olin   xi

Preface
Penny Travlou and Catharine Ward Thompson   xvii

Part 1 POLICY ISSUES: WHAT ARE THE CURRENT CHALLENGES IN PLANNING FOR INCLUSIVE ACCESS?   1

1 Public spaces for a changing public life
Jan Gehl   3

2 'The health of the people is the highest law': public health, public policy and green space
Ken Worpole   11

3 Playful nature: what makes the difference between some people going outside and others not?
Catharine Ward Thompson   23

Part 2 THE NATURE OF EXCLUSION: WHAT IS THE EXPERIENCE OF EXCLUSION IN DIFFERENT CONTEXTS?   39

4 Culture, heritage and access to open spaces
Judy Ling Wong   41

5 Landscape perception as a reflection of quality of life and social exclusion in rural areas: what does it mean in an expanded Europe?
Simon Bell and Alicia Montarzino   55

6 Mapping youth spaces in the public realm: identity, space and social exclusion
Penny Travlou   71

CONTENTS

Part 3 DESIGN ISSUES: WHERE ARE THE DESIGN CHALLENGES AND WHAT DOES INCLUSIVE DESIGN MEAN IN PRACTICE? 83

7 What makes a park inclusive and universally designed?: a multi-method approach
Robin C. Moore and Nilda G. Cosco 85

8 'You just follow the signs': understanding visitor wayfinding problems in the countryside
Katherine Southwell and Catherine Findlay 111

9 Developing evidence-based design: environmental interventions for healthy development of young children in the outdoors
Nilda G. Cosco 125

10 Healing gardens for people living with Alzheimer's: challenges to creating an evidence base for treatment outcomes
John Zeisel 137

Part 4 RESEARCH ISSUES: WHERE ARE THE RESEARCH CHALLENGES AND WHICH THEORIES AND METHODS OFFER MOST PROMISE? 151

11 Measuring the quality of the outdoor environment relevant to older people's lives
Takemi Sugiyama and Catharine Ward Thompson 153

12 Three steps to understanding restorative environments as health resources
Terry Hartig 163

13 On quality of life, analysis and evidence-based belief
Peter A. Aspinall 181

Index 195

List of contributors

Peter Aspinall is Associate Director of OPENspace Research Centre and holds a Masters degree in Psychology and a PhD inthe Faculty of Medicine at Edinburgh University. He has undertaken teaching and research in Architecture and Landscape Architecture at Edinburgh College of Art. He was appointed Research Director of Environmental Studies, and Research Director of the College before moving to the School of the Built Environment at Heriot-Watt University. In addition to being an Associate Director of OPENspace, he is co-founder of VisionCentre3, a new collaborative research centre for visually impaired people, and has introduced new courses for disabled people on inclusive access and environmental design. He is currently an Honorary Fellow of Edinburgh University and Emeritus Professor of Vision and Environment at Heriot-Watt University.

Simon Bell is Associate Director of OPENspace and a forester and landscape architect, educated at University of Wales, Bangor and Edinburgh. He is an international expert on forest and park landscapes and large-scale landscape evaluation and design. His research interests include landscape and recreation planning and design, and he is the author of several books published by Taylor and Francis. He has undertaken many projects in Britain, Canada, the USA, Ireland, Latvia and Russia. He is also an Associate Professor in the Department of Landscape Architecture at the Estonian University of Life Sciences.

Nilda G Cosco is the Co-ordinator of the Initiative for Inclusive Design, Research Associate Professor and Education Specialist at The Natural Learning Initiative (a research, design assistance, professional development, training and dissemination unit of the College of Design), North Carolina State University. Dr Cosco holds a degree in Educational Psychology, Universidad del Salvador, Buenos Aires, Argentina and a PhD in Landscape Architecture, Edinburgh College of Art/Heriot-Watt University, Scotland. She has an interest in the impact of outdoor environments on health outcomes such as obesity, sedentary lifestyles, attention functioning and wellbeing. Her current research is supported by the US National Institute of Environmental Health Sciences (NIEHS), the US National Science Foundation (NSF), and the Robert Wood Johnson Foundation. Dr Cosco is

LIST OF CONTRIBUTORS

co-author of the *Child's Right to Play in Large Cities* (in Spanish), 'Our Neighborhood is Like That!', in *Growing Up in an Urbanizing World.* (Chawla, ed., 2002) (Earthscan, 2002) and *Well-being by Nature: Therapeutic Gardens for Children* (American Society of Landscape Architects, 2006).

Catherine Findlay is an honorary research fellow in OPENspace and a qualified occupational therapist. Recent projects include: Wayfinding and Visitor Information (Forestry Commission), Access to the Countryside for Deaf Visitors (Scottish Natural Heritage), Places for People and Wildlife in Cumbria (English Nature), Inclusive Interpretation (The Eden Project/Sensory Trust) and An Evaluation of Therapeutic Horticulture (Thrive).

Jan Gehl is an architect, Professor of Urban Design and Director of the Centre for Public Space Research at the School of Architecture, Royal Danish Academy of Fine Arts, Copenhagen. He is also founding partner of GEHL Architects ApS – Urban Quality Consultants, Copenhagen. His international profile includes teaching at universities in Edinburgh, Toronto, Calgary, Melbourne, Perth, Berkeley, San Jose, Oslo, Dresden, Wroclaw, Hanover, Guadalajara, Vilnius, Cape Town and Costa Rica. Consultancies include city centres in London, Edinburgh, Melbourne, Adelaide, Perth, Wellington, Cape Town, Amman, Zurich, Riga, Oslo, Stockholm and Copenhagen. Publications include *Life Between Buildings – Using Public Space*, published in fifteen languages, *Public Spaces, Public Life, Copenhagen*, winner of the EDRA/PLACES Research Award, USA, 1999, and *New City Spaces*, Danish Architectural Press, Copenhagen, 2001, published in six languages. Jan Gehl has been awarded the Sir Patrick Abercrombie prize for exemplary contributions to town planning by the International Union of Architects as well as an honorary doctorate from Heriot-Watt University, Edinburgh.

Terry Hartig is Associate Professor of Applied Psychology at the Institute for Housing and Urban Research and Department of Psychology at Uppsala University, Sweden. He also holds an Adjunct Professorship in Environmental Psychology with the Departments of Plant and Environmental Sciences and Landscape Architecture and Spatial Planning at the Norwegian University of Life Sciences. He uses experimental and epidemiological methods in efforts to assess and explain the health resource values of experiences in natural environments.

Alicia Montarzino is a lecturer at the School of the Built Environment, Heriot-Watt University and a research fellow in OPENspace. She is an architect and landscape architect, with a Masters degree in urban and regional planning and a PhD in environmental psychology. She worked as a Director of Public Works and hospital designer in Argentina before becoming an independent research consultant in Sweden. She has also been involved in research at the University of Puerto Rico and at the University of Illinois at Urbana-Champaign. Her current areas of research are perceptions of open spaces and visual impairment and perceptions of safety in the build environment. She is currently the director of Inclusive Environmental Access and Design, a CDP course that Heriot-Watt University offers to train disabled students as environmental access auditors.

Robin Moore is Professor of Landscape Architecture and Director of the Natural Learning Initiative, North Carolina State University. Professor Moore is an urban designer and design researcher, specialising in child and family urban environments. He previously taught at the University of California, Berkeley and Stanford University. He is a member of the UNESCO Growing Up in Cities international research team, with particular involvement in Buenos Aires, Argentina and Amman, Jordan, projects. He holds degrees in architecture (London University) and city and regional planning (Massachusetts Institute of Technology). His publications include *Natural Learning* (1997); *Plants For Play* (1993); *The Play For All Guidelines* (1987, 1992) and *Childhood's Domain* (1986).

Laurie D. Olin is Practice Professor of Landscape Architecture and Regional Planning at the University of Pennsylvania and founding partner of the Olin Partnership, Philadelphia. He has won the Bradford Williams Medal for his writing on the history and theory of landscape architecture and is a Fellow of the American Academy of Arts and Sciences as well as the American Academy of Arts and Letters.

Katherine Southwell is an honorary research fellow at OPENspace Research Centre, Edinburgh College of Art/Heriot-

LIST OF CONTRIBUTORS

Watt University. Her research interest is in user experience (of both indoor and outdoor environments), and in the application of observational research methods for environment-behaviour studies. Katherine holds a PhD in Landscape Architecture.

Takemi Sugiyama is a research fellow at the Cancer Prevention Research Centre, School of Population Health, the University of Queensland, Australia. Dr Sugiyama is currently working on various research projects concerning environmental determinants of physical activity and sedentary behaviour. He holds a Bachelor of Engineering (Nagoya), Master of Architecture (Virginia Tech), and a PhD in Environment-Behaviour Studies (Sydney).

Penny Travlou is qualified in sociology and cultural geography, and previously worked with the National Technical University of Athens and Edinburgh University. Her research interests lie in the field of cultural/urban geography. She holds a PhD from the Department of Geography, University of Durham. Since 2002, she has been working as a research fellow at OPENspace Research Centre (Edinburgh College of Art/Heriot-Watt University), including a project on the perception and use of public open spaces by teenagers in Edinburgh. For this research, she was awarded an Early Career Research Grant (Edinburgh College of Art), a Small Research Grant (The Carnegie Trust) and, with Professor Ward Thompson, a Joint Activities Award (the British Academy). She has presented her research work at numerous international conferences. She is vice-chair of the International Play Association in Scotland.

Catharine Ward Thompson is Research Professor of Landscape Architecture at Edinburgh College of Art. She is Director of OPENspace – the research centre for inclusive access to outdoor environments – based at Edinburgh College of Art and Heriot-Watt University and directs the College's Landscape Architecture PhD programme. She is also Director of a multi-disciplinary research consortium entitled I'DGO (Inclusive Design for Getting Outdoors), which focuses on quality of life for older people, in collaboration with the Universities of Salford, Oxford Brookes, Heriot-Watt and Edinburgh. She is an active member of SPARColl (the Scottish Physical Activity Research Collaboration) led by the University of Strathclyde. She is a Fellow of the Landscape Institute and the Royal Society for the encouragement of Arts, Manufactures and Commerce (RSA).

Judy Ling Wong is the Director of the Black Environment Network. BEN works across sectors, integrating social, cultural and environmental concerns. Judy is a major voice on social inclusion policy, has worked extensively in various sectors and is uniquely placed to lead an integrated approach to environment participation.

Ken Worpole is a freelance writer and environmentalist, and a member of the UK government's Urban Green Spaces Task Force. He is an adviser to the Commission for Architecture and the Build Environment and the Heritage Lottery Fund. His most recent book, *Last landscapes: the architecture of the cemetery in the West*, was published by Reakton Books in 2003. Details of other publications can be found at: www.worpole.net

John Zeisel is Visiting Professor, School of the Built Environment, Salford University and president and co-founder of Hearthstone Alzheimer Care, Ltd, Massachusetts. He is the author of *Inquiry by Design: Environment/Behavior/Neuroscience in Architecture, Interiors, Landscape and Planning*, W.W. Norton, 2006

Foreword

Laurie Olin

Recently in a museum, I found myself looking at a chair made 3,000 years ago in Egypt. It was elegant in its proportions and details. It also looked like it could have been made yesterday, and that it would be very comfortable to sit in. Nearby were a pair of gloves from the same era, and other daily items of life, including several loaves of bread that could have been smuggled in from a bakery a few blocks away, if it weren't for the fact that they had been stale for thousands of years. How timeless these human products were. The continuity of human life, needs and accommodations was forcefully apparent. How like us these ancient people seemed.

And yet, how unlike us they were: the world-view and aspirations of the majority of this ancient population had almost nothing in common with us today, in a cultural sense. While the wheat they grew and cotton they made were uncannily similar to that which is produced in the same river valley today, there were no automobiles, freeways, electronically amplified music and calls to prayer, no high-rise apartment buildings with neon signs, no movies and televisions bringing news and MTV from distant parts of the world, along with alternative life-styles. In short, despite how much we have in common with people from another time, we also are significantly different in how we live, think, work and play. As one travels about the world, it also becomes apparent that, despite the many things that are shared among the world's populations today, there are radical differences between the inhabitants of different countries, even those with a common cultural history. We only have to open the newspaper each day to read of the tragic events of the moment that are partially the result of such differences, of misunderstanding and conflicts in perceived needs, desires and values.

It seems almost axiomatic that, even while remaining gregarious primates with archaic physiological needs and aspects of behaviour, on the one hand, and having developed highly evolved and constantly changing cultural and social beliefs, values, needs and desires on the other, there would be some, even if only a small handful, eternal lasting truths about human environmental design. It would also seem that some things appropriate in one age or location would absolutely be suitable in one age and not another.

The thoughtful and frequently provocative, even at times counterintuitive, essays gathered here support such conclusions, but go far beyond such generalities. Their value lies in

FOREWORD

particularities and specific insights, more than in sweeping conclusions. Here we find clear, even bold conclusions, in many cases worthy of dissemination and absorption by anyone engaged in physical planning, designing, commissioning or administering portions of our physical world, whether urban or rural.

Thoughtful individuals have considered the planning of cities and the design of buildings and open space since classical antiquity, and even earlier in the ancient near east. Marcus Vitruvius Pollio, writing in the first century BCE, summarized a considerable amount of thought on the subject of planning and design in his *Ten Books of Architecture*. Among many sensible remarks regarding the orientation and planning of urban streets and buildings he remarked:

> The space in the middle, between the colonnades and open to the sky, ought to be embellished with green things; for walking in the open air is very healthy, particularly for the eyes, since the refined air that comes from green things, finding its way in because of the physical exercise, gives a clean-cut image, and by clearing away the gross humours from the frame, diminishes their superabundance, and disperses and thus reduces that superfluity which is more than the body can bear.
>
> (Vitruvius, Book 5, Chapter 10, paragraph 5, translated by Morgan, 1914)

While the language and explanations are different, Frederick Law Olmsted, writing in the nineteenth century in America, and many of the authors in this text also, believed and set out to find why and how fresh air, natural elements and open space of particular sorts can enhance life and health. What exactly are the facts that can explicate views that seem so obviously, but mysteriously, true?

Vitruvius famously grouped many topics of design under three general headings that have been translated as *firmness*, *commodity* and *delight*. In our own time, J. B. Jackson led many in my generation to a reformulation of this triad in a remarkable essay, 'The Imitation of Nature':

> As a man-made environment every city has three functions to fulfill: it must be a just and efficient social institution; it must be a biologically wholesome habitat; and it must be a continuously satisfying aesthetic-sensory experience.
>
> (Jackson and Zube, p. 87)

To acknowledge that few of our cities live up to these aspirations is to acknowledge that there is an awful lot to do on the part of those who are involved in their design and planning. Landscape Architecture and Architecture are distinguished from what are often called 'fine' arts by the fact that they are considered 'useful' arts. As such, they need to answer to the needs of society and the individual, a difficult task given how diverse our populations have become. Such demands are not unreasonable or impossible, as has been amply demonstrated by the innumerable successful villages, towns, buildings and landscapes that have existed throughout history on every continent (except Antarctica).

It is also true that, throughout history, there have been poorly planned, badly built, unpleasant, dangerous and outright disastrous settlements, structures, cities and landscapes. The twentieth century has experienced widespread examples of both the best and the worst of these. To the dismay of many, some of the dullest, meanest, most ill-conceived, poorly designed and hostile human conditions to have been planned, designed and built have been created in the past century in the United States and Europe, where governmental institutions, economic systems and professional education and employment are thought to be among the most developed and effective in the world. How could nations that have so many trained professionals have produced such banal, dysfunctional, unsupportive, even dangerous, environments?

In part, it has to do with the cycles of interest and fashion on the part of society and designers confronting Vitruvius' topics, of emphasizing one over the others and, worse, in part from ignorance regarding human needs and behaviour. Looking at the profession of architecture, it is clear that in the early years of the last century, a great emphasis was placed upon 'firmness', on new structures, methods and materials for building, upon innovations that radically altered multistorey buildings throughout the world, changing the face and nature of cities, with an increased ability to produce density and environmental conditions different from those of earlier eras.

FOREWORD

The crowding and deprivation of resources, both natural and societal, insufficient infrastructure and the concomitant physical and mental health problems that have resulted are well known. They were repeatedly addressed by Olmsted, Geddes and Mumford (as well as by many others), each of whom thoughtfully studied urban situations. All proposed various landscape and open space remedies to counteract the damage and effects upon the citizens of cities transformed by the latest architecture, transportation and working conditions.

In the decades between the two great wars of the past century, and in the reconstruction and economic situations of rebuilding after each of them in North America and Europe, considerable emphasis in the planning and design community was placed upon 'commodity', as it was understood at the time, with the assumption that other important issues and needs would fall in line and be achieved as a result. The adage 'form follows function' became a mantra, but a full appreciation of 'function' seems to have eluded many practitioners and policy makers. The greatly increased quantum of urban development and a generation of socially progressive governments and designers educated under a machine-age aesthetic and functionalist banner, combined with the difficulties of financing such a volume of construction amid Cold War politics and political manoeuvring, was brought to an abrupt halt by the widespread events of 1967 and 1968, on both sides of the Atlantic. Outrage with local and foreign policies, with disastrous wars, with racial, ethnic and sexual inequity, with failed political and economic policies, led to riots, protests and upheaval in the United States, France, Germany, Italy, Czechoslovakia, the Middle East, Mexico and the United Kingdom.

Architects, landscape architects and planners were in the thick of such events in all of these countries, in some cases leaders of opposition to established methods and authority. One result was that considerable attention was given to theory and research in design and planning throughout the 1970s as young designers, academic institutions, social and natural scientists focused upon the problems that had emerged. Rachel Carson's *Silent Spring*, Jane Jacobs' *Death and Life of Great American Cities*, William H. Whyte's *The Exploding Metropolis* and *The Last Landscape*, Oscar Newman's *Defensible Space*, Constance Perrin's *With Man in Mind* and a host of other books, articles, lectures and events challenged designers and planning agencies to rethink their ideas and habits. One result was the creation of various groups and organizations such as EDRA, the Environmental Design Research Association, which endures today, and which began studying and disseminating emerging original research about what was really going on in the world, how people really used open space and behaved, what they truly thought and needed, and how designers could produce more efficacious places.

Brilliant individual figures emerged as leaders, with new methods and results, such as landscape architects Lawrence Halprin and Ian McHarg. Halprin developed a format for community workshops that focused upon collectively establishing programmes and site design strategies that were agreed upon to serve the needs and aspirations of a particular community and place. McHarg developed an overlay method of analysis that considered a full panoply of natural and several social factors that could be used to produce matrices and alternative syntheses for landscape planning, whether for a particular development, preservation, or a given set of values and goals. Many planners and designers throughout the developed world were stimulated by these developments and became engaged in similar activities and methods. One positive result was the creation in the United States, Canada and Britain of environmental legislation, regulation and agencies at various levels of government to oversee improved land planning and design.

From the standpoint of 'commodity', a lot of good work took place on both sides of the Atlantic in the 1970s and 1980s. The profession of landscape architecture as a whole made great strides in general knowledge, attitudes and methods of land planning in terms of the biophysical realm. On the social side of the ledger, the results were more mixed and less grand. Several of America's leading academic landscape departments attempted to engage social science. Both the University of California at Berkeley and the University of Pennsylvania in Philadelphia placed social scientists (as well as natural scientists) within their faculties of landscape architecture. While it is undoubtedly true that numerous teachers and practitioners emerged from these and other institutions, inspired by, and to some degree more sensitive to, issues regarding the social use of space, it is also obvious that the practice of landscape architecture has not become particularly enlightened or been

FOREWORD

transformed in the same way that it has regarding ecological issues. There are several probable reasons for this. First, the basic knowledge, the fundamental data, explication and theory regarding human ecology and sociology (let alone the physiology) of human–environmental interaction is nowhere nearly as developed, understood or agreed upon as that of natural science. Second, applied science in the hands of others, whether it be those of politicians or designers, requires great skill and care, knowledge and insight. As recently demonstrated by the issue of global warming and attempts to persuade leaders in developed countries such as China and the United States to make changes regarding emissions, energy and the carbon cycle, even with an overwhelming body of evidence regarding our best interests, it can be very difficult to change current development practices.

Social science ventures into even more hotly contested territory. At issue, almost always, are questions about the values and potential biases of those conducting the research and their motives. The ghost of discredited political and social agendas that had disastrous results in the twentieth century haunts our age. And yet, to be ignorant of facts, to avoid research and to continue to attempt to produce homes, communities, cities and regions employing only intuition and the experience and politics of a few individuals, no matter how well meaning, intelligent and representative of the majority of a group, can be woefully inadequate and, at times, verge on the criminal. The design and planning fields today simply need more and better information to base our work upon.

The landscape faculty at Penn, for a time, dreamed that a sequel to *Design With Nature* could be developed. *Design With Man* never happened as debates about determinism, class, inequity and conflicting social values erupted in studios, design workshop reviews and faculty meetings. The social scientists were often dismayed by the questions and pre-conceived interviews the designers developed. The designers were consistently troubled that the social scientists were oblivious to deeply felt aesthetic or ecological concerns. Ultimately, whatever fruit such collaborations might have yielded was not to be in that decade, for the political climate in the United States and the United Kingdom, and several other European countries, removed the financial support that had allowed them to begin in the first place.

By the early 1980s, a conservative backlash against the optimistic liberal agenda of the 1950s, 1960s and 1970s had begun. The governments of Ronald Reagan and Margaret Thatcher dismantled agencies, public utilities, regulations and portions of the environmental measures only recently established, and cut financing for research in areas of particular interest to the field of landscape architecture, especially regarding the environment, energy and public health.

It is not that nothing happened after these developments. In 1980, William H. Whyte published the *Social life of Small Urban Spaces*. For several decades, Erv Zube and his colleagues at the University of Massachusetts produced a series of research studies on public attitudes and perceptions regarding landscape. *Landscape Journal* was founded to disseminate scholarly articles, particularly social research and theory. Elsevier in Europe sought and published research. EDRA continued with annual meetings, publications and awards. Clare Cooper Marcus, Randy Hester and their colleagues at Berkeley, Jay Appleton, Yi-Fu Tuan, Bill Hillier and others at the Bartlett School in London, Catharine Ward Thompson in Edinburgh, Jan Gehl in Copenhagen, and many others have continuously issued work in the past two decades that has looked at the design of open space and human behaviour, asking what do people do? Why? What do they think about their spaces, their lives and their quality? What works and what doesn't? The range of philosophy, theory, experiments, studies, findings and statistics has been very rich. And yet . . .

As the tide of funding for research on the part of genuine scientists was receding, but not particularly related to it, an upsurge in the third topic of Vitruvius, that of 'delight', welled up in landscape design circles, in the unlikely forms of postmodern classicism, eclecticism and land art. The long years of supposedly 'functional' planning and buildings had produced so much that was dull, depressing and banal at best, and most certainly poverty stricken, artistically and spiritually. Many landscape architects turned to the past, both atavistic and historical, for inspiration. As can be witnessed in the pompous, stultifying neoclassical work produced in Germany, France, Britain, the United States and Canada in the late 1980s and 1990s, even less thought was given to the needs and intelligence of children, the elderly, minorities or even the middle-aged, lower and middle-class workers paying for and doing much of the work of society.

FOREWORD

Likewise, many landscapes made in this period by designers yearning to make 'art' with Bronze Age forms, space age materials and collagiste organizational strategies have, as often as not, proved to be as unusable, dysfunctional and unloved as the vapid historicist schemes of the period.

It is understandable what prompted some of this work, even while regretting the approach or methods employed. One of the 'functions' that a landscape can have, and which can contribute greatly to its preservation and sustainability, is 'Beauty'. While definitions of what comprises beauty have varied from place to place and from one society to another over time, there seems to be no question that each society has a sense of this phenomenon. Several of the studies that follow in this collection allude to this in their findings. Whether it is Alzheimer's patients in managed care or urban families on a weekend outing in the countryside, the search for pleasure afforded by natural elements shines through. Likewise, the beneficial results to be had through contact with natural elements, especially in the heart of our cities, are verified again and again in several of the chapters. One virtue of essays such as these is their verification with careful observation, facts and figures of several deeply held beliefs of landscape architects, namely that many, if not most, humans have an inbuilt need for natural scenery and particular aspects of it, specifically: water, trees, flowers, sunlight and earth. This is as true for those in the best of health as it is for patients recovering from illness and surgery.

By its very nature, landscape architecture stubbornly remains as much of an 'art' as a science. This is partly because no matter how much we know about human history, behaviour and natural science, every project is, in one way or another, a full-size experiment in real time and space. It has never been built there before, at this time of year, for these particular individuals. But our accumulating knowledge can help us to make informed assessments of what people might feel as they walk into a space of a particular size and proportion, of how a particular species of tree will fare in a particular climate and exposure, how things will physically hold up or break down. We know that certain verities will remain, that people will take comfort from shade when it is hot and pleasure from sun when it is cool, that the sound of splashing water can be both soothing and stimulating, and that people, especially children, will be drawn to it. We also know that most humans like to claim space by adjusting it in some way to themselves, whether it be shifting a chair slightly before sitting in it to watch or engage others, or digging holes and building mounds if one is a child.

The following chapters range across several areas of such investigation, prompting conclusions regarding health, behaviour and how environments – through their organization or arrangement, their physical attributes and elements, resulting from their planning or design – promote individual and community health and social wellbeing. They look at topics as widely diverse as wayfinding systems and signs in rural scenic areas, differences in landscape perceptions for immigrant and long-established communities, the effects of particular urban housing layouts upon older people's mobility and teenage aggression patterns. The essays here confirm these rudimentary truths, but go so much further in their particular investigations; they contain material that every professional should take in.

Those who have taught know all too well that, although most designers are interested in ideas, like to read and have a great desire to know and understand as much about their world as possible, it can be very difficult at times to get students (or practitioners, for that matter) to step away from their computers and drawings, from their long sessions trying to solve problems, to quietly read. As teachers, we also yearn for material that will engage them, feed them and help inform their work. We cast about for literature that can help them grow and move beyond personal whim and the limits of their own personal experience and habits. It is essays and studies such as those in this collection that can help.

Not only do we have these studies, but also each author offers a useful bibliography on their topic, in several cases, helpfully supplying entries on their own, often extensive, previous work. This collective bibliography alone makes this a book that should enter every professional landscape architect's office. Over the years as a professional, I have cast about for the documentation to help give clients and public figures assurance that many of the old truths and common-sense notions that we have offered them are not merely personal whims or speculative and wishful thinking. Again, it is work such as is to be found here that can help fill this void. It is also the sort of work that prompts one to ask and hope for more to come from each of these talented and curious souls.

FOREWORD

References

Jackson, J. B. and Zube, E. H. (1970) *Landscapes: Selected Writings of J. B. Jackson*. Amherst: University of Massachusetts Press.

Vitruvius, in Morgan, Morris Hickey (1914) *The Ten Books on Architecture*. Edited by Tom Turner in 2000. Available at www.lih.gre.ac.uk/histhe/vitruvius.htm.

Preface

This book arose out of a desire to debate current theory and practice in the planning, design and management of inclusive access to the outdoor environment. Our aim was to collect in one volume a range of perspectives on what constitutes good design for socially inclusive open space, and what research there is to support this, at a time when conceptions of indoors and outdoors, public and private, collective and individual space and its uses are being redefined under the pressure of early twenty-first-century anxieties and hopes. The themes of the book reflect the developing arena in which OPENspace research centre has been a key player since its establishment in 2001. We have invited contributions from international leaders in their field, who bring a range of disciplinary perspectives – from urban planning, architecture and landscape architecture to public policy, environmental psychology and urban and cultural geography – to the analysis of policy and planning, inclusive design and innovative research directions focusing on people's access to and engagement with open space.

This book emerges from the belief that inclusive access to high-quality public spaces is a cornerstone of democracy and social equity, a fundamental condition for social and political participation, and a key element with potential to enhance wellbeing and quality of life. Inclusive access to open space should, thus, be seen as central to good planning and design practice. Indeed, governmental interventions that aim to improve people's quality of life in Britain, across Europe and in many other parts of the world have recently tended to prioritise, in declarations if not always in practice, some kind of design for inclusion. Issues of inclusive access are at the heart of programmes for urban renaissance and for revitalising the tourist industry in town and countryside. In the United Kingdom, implementation of equal opportunities, race relations and disability discrimination legislation has emphasised the need to widen access to goods and services for all people and to address users' needs directly. In these contexts, and many more besides, access to open space and the design that facilitates it are deemed indispensable. But what is meant by '*design for inclusive access*'? How can it be implemented in the variety of contexts and situations that call for attention? How can methods and practices in inclusive design be refined by the insights that theory and research offer in attempting to understand the variable contexts of social exclusion and the

PREFACE

way people perceive, use and respond to open space? And what are the challenges, and possible limitations, of these research-inspired understandings? This book contextualises design for inclusive access by addressing detailed aspects of access to, and engagement with, natural and outdoor environments for people of different ages, abilities, ethnicities and socio-cultural groups. Presenting empirical research on the use of open space in a range of social and environmental settings, the book and its contributors reaffirm the importance of access for all to environments that are rich in opportunities and support for health, development and wellbeing. Indeed, it underlines the thesis that open space should be a place of delight and pleasure, eliciting and responding to the 'playful natures' in all of us.

The book is organized into four sections. First, it covers policy issues in planning for the interaction between people and the outdoor environment at all scales and in both urban and rural contexts; second, it explores the experience of exclusion in distinct contexts and from the perspective of race, age and culture; third, it focuses on the challenge of evidence-based design for inclusive access to outdoor places; and, finally, it offers insights from those involved at the forefront of research on innovative approaches and new understandings in the field.

Jan Gehl opens the first section by reminding us of the importance of designing public spaces for a changing public life. Drawing upon examples from his own professional experience over several decades in Copenhagen, he presents a summary of the arguments that underlie the renaissance of public space and public life in twenty-first-century European city culture, a renaissance in which he has been a pioneer. In Chapter 2, Ken Worpole draws together the agenda of planning for inclusive access to green space and public health policy. He argues that policy makers should acknowledge the role that informal outdoor recreation has always played in public health, and reflect this in current policies for investment. In Chapter 3, Catharine Ward Thompson closes the first section by exploring how people engage with their outdoor environment and presenting evidence for some of the parameters that influence the ways people perceive and use open space.

The second part of the book looks at different facets of social exclusion. In Chapter 4, Judy Ling Wong reminds us that involvement of ethnic communities is one of the key challenges in planning and designing accessible open spaces. She shows how community involvement allows open spaces to be created and managed in socially and culturally relevant ways. In Chapter 5, Simon Bell and Alicia Montarzino investigate aspects of social exclusion experienced by people living in contrasting rural areas. Using Scotland and Latvia as case studies, they explore the changing relationship between perceptions of socio-economic disadvantage, physical environment and quality of life among different generations in the Europe of the twenty-first century. Finally, in Chapter 6, Penny Travlou investigates what teenage-friendly open space might mean. She suggests that designers need to understand the uniqueness and complexity of young people's use and experience of public space, often imbued with tensions and challenges, but still fundamental for their development and wellbeing.

The third part of the book explores inclusive design in practice and the accumulating evidence to support certain approaches and design solutions. In Chapter 7, Robin Moore and Nilda Cosco challenge designers to provide high-quality public spaces and utilise a multi-method approach to assess social inclusion in a USA park developed on universal design principles. Their results offer eminently practical insights in environment/behaviour dynamics for planners and designers wanting to provide inclusive park environments. In Chapter 8, Katherine Southwell and Catherine Findlay have analysed wayfinding challenges as a barrier to accessing the countryside and present the innovative Site Finder assessment toolkit as a vehicle for translating theory into practice to make wayfinding infrastructure more effective. In Chapter 9, Nilda Cosco focuses on the design of outdoor play environments to support healthy development in young children. She argues that childcare outdoors is the strongest correlate of physical activity in preschool children and proposes the use of evaluation instruments such as POEMS (Preschool Outdoor Environments Measurement Scale) to measure the overall outdoor quality of preschool play areas. In Chapter 10, John Zeisel engages with the opportunities and challenges of inclusive design in relation to cognitive impairment, drawing on examples of healing gardens for people living with Alzheimer's. He argues that, while the theory and practice of healing garden design is quite advanced, post-occupancy evaluation still presents us with significant methodological challenges.

PREFACE

The fourth part of the book focuses on some of the most promising and innovative theoretical and methodological approaches to researching inclusive access to outdoor environments. In Chapter 11, Takemi Sugiyama and Catharine Ward Thompson explore meaningful ways to understand the relationship between the physical environment and older people's levels of activity. Their research develops the concept of 'environmental supportiveness' and two instruments for measuring the quality of the environment relevant to older people's access outdoors and quality of life. In the twelfth chapter, Terry Hartig explores some of the ways that restorative environments, such as parks and open spaces, promote mental and physical health and wellbeing. He provides a lucid account of the complex theories behind this concept and analyses three methodological approaches to the research of restorative environments: study of discrete restorative experiences; study of cumulative effects of repeated restorative experiences; and study of social-ecological influences on access to, and use of, places for restoration. The thirteenth and final chapter of the book, by Peter Aspinall, raises the question of how quality of life can be assessed in social research and offers some unusual and innovative quantitative methods for use in landscape architecture and environmental design. He finishes by describing an unconventional Bayesian approach to illuminate, particularly in controversial circumstances, what is believable from typical research findings in the field.

In brief, this book is about people's access to outdoor environments – streets and squares, gardens and parks, woodlands, and the wider countryside. It reviews recent evidence about the nature and value of people's experience of such open space, and analyses what is important in good design to meet people's needs and desires in the twenty-first century. It looks to the future and suggests innovative ways to develop an inclusive understanding of how the landscape, urban or rural, can contribute to health and quality of life.

The book's contents reflect the developing body of expertise accumulated by OPENspace Research Centre and the insights gained from the *Open Space: People Space* conference hosted in Edinburgh in 2004. This landmark conference brought together over 250 delegates from around the world, multidisciplinary in scope and extending into the domains of policy and practice as well as those of academic research. The breadth of expertise at the conference is reflected in the authors of this book, offering a rare insight into people's engagement with the outdoor environment at all scales and in relation to a range of themes. It will be of value to policy makers, researchers, designers, planners and managers.

We hope the book will be of particular interest to the design professions – landscape architects, urban designers, architects – as well as planners, social and environmental scientists, health policy makers and professionals, and those working in the social services, including child development, health care and community development. It will also be a useful tool for students who would like to get an in-depth understanding of inclusive access to outdoor environments, looking at conceptual and practical aspects of social inclusion and the methodological and theoretical challenges of implementing inclusive design.

This book would not have materialised without the patient and generous co-operation of the contributing authors and the valuable time and effort expended by Peter Aspinall and Simon Bell, who assisted with reviewing the texts. We thank you all. We are particularly grateful to Anne Boyle for her exemplary and good-humoured editorial assistance; without her this book would have never met its deadlines and might well have lost a few authors on the way. We are also grateful to Anna Orme who, as OPENspace administrator, has managed the whole, long process from original conference concept. We thank the designers and photographers who allowed us to use copyrighted visual material in this book. Finally, we are grateful to our publisher, Caroline Mallinder, who supported the concept of the book from the start, Kate McDevitt, our commissioning editor, Jane Wilde, our editorial assistant, and Stephanie Kerrigan, our production editor, at Taylor and Francis.

PENNY TRAVLOU AND CATHARINE
WARD THOMPSON, FEBRUARY 2007.

Part 1

Policy issues: what are the current challenges in planning for inclusive access?

Public spaces for a changing public life

Jan Gehl

Introduction

Recent decades have seen a gradual development from industrial society's *necessary public life* to the *optional public life* of a leisure and consumer society. Where city life was once a necessity and taken for granted, today it is to a high degree optional. For that very reason, this period has also seen a transition from a time when the quality of city space did not play much of a role in its use, to a new situation in which quality is a crucial parameter. In the past, people had to use the streets and squares of the city regardless of their condition. Today this is in the majority of cases an option.

It should, at this point, be noted that the changes described in this chapter relate to societies where the economic developments have initiated a shift towards leisure and consumer-oriented lifestyles. In many less developed regions in the world, life in public spaces typically has not changed materially to such a degree and is still found to be very much dominated by necessary activities.

The traditional city: meeting, market and moving

Seen in a long-term historical perspective, city space has always served three vital functions – meeting place, marketplace and connection space. As a meeting place, the city was the scene for exchange of social information of all kinds. As a marketplace, the city spaces served as venues for exchange of goods and services. And finally, the city streets provided access to and connections between all the functions of the city (Gehl and Gemzøe, 2001).

This pattern can be followed from the earliest urban settlements through Greek and Roman cities, medieval cities, renaissance and baroque cities as well as cities from the age of enlightenment and the industrial age. City spaces have teemed with people and functions throughout history. Life in city space was an integral and utterly essential part of society. Numerous descriptions, paintings and engravings from various periods in history, as well as pictures from the early days of photography, vividly tell this story.

JAN GEHL

In the medium term, referring to situations just a hundred years ago, the patterns continued, as seen in photographs and street scene engravings from around 1900. City spaces functioned as meeting place, marketplace and connection space. The streets were crowded with people burdened with goods and packages or simply on their way by foot through the city, a testimony to a time when few other forms of transport were available. Goods were sold from booths or by street peddlers. Numerous people of all ages were on the streets and squares to take part in city life or simply because there was not enough room inside their crowded dwellings, small shops and cramped workshops. It is clear from old pictures that public city life was completely dominated by the activities essential to everyday existence.

Twentieth century: a near farewell to public life

In the course of the twentieth century a number of important developments have taken place in cities such as Copenhagen. First and foremost, these hundred years have seen a dramatic improvement in the economic conditions of the city and its dwellers. This development accelerated, especially in the second part of the century. The growth in the economy led to a wide range of changes in society situation and in lifestyles. Two developments, derived from the changes in economic conditions, have had a particularly dramatic influence on the concept of public spaces as well as on their physical appearance and the conditions for public life.

The Modern Movement, from the mid 1920s onwards, in its quest to provide growing urban populations with cleaner and healthier cities and accommodations, dramatically downgraded the importance of traditional public spaces. Streets and squares were declared unhealthy and unwanted, and activities in such places were severely criticized as being shady, unbeneficial and, to a wide extent, amoral. Parkland settings for housing, with trees and lawns as meeting places instead of streets and squares, would be the new answer to the calamities of the traditional townscapes. The CIAM Athens charter of city planning (1933) laid down the new rules and stated that residences, work, recreation and transport should be strictly separated in the modern city. This dramatic ideological condemnation of traditional forms of public space and public life would, for several decades, effectively stop any development of the townscape as well as research and discussions concerning public life.

The other dramatic development, which drastically changed the conditions for cities, public spaces and public life in the twentieth century, was the influx of motor cars in great numbers. The car invasion had been going on at a moderate rate since the beginning of the century, but really took off in the mid 1950s, some ten years after the Second World War.

Thus, by the early 1960s, a situation had developed where new Modernistic planning concepts had more or less phased public life out of the new city districts, while in all the older parts of the cities, what remained of public life was harassed or simply squeezed out of streets and squares by traffic and parking.

Early 1960s: a turning point

The book, *The Death and Life of Great American Cities*, published in 1961 by Jane Jacobs, marked a turning point in the gradual erosion of the concept of public spaces and public life.

1.1 1880. Old engraving showing city life on Strøget, Copenhagen's main street, a century ago. Essential, work-related errands dominate.

PUBLIC SPACES FOR A CHANGING PUBLIC LIFE

1.2 c.1960. Strøget invaded by cars. City life is forced onto narrow pavements with room for only the most essential pedestrian activities.

1.3 1968. Strøget five years after conversion to a pedestrian promenade. Car-free streets are seen as the antidote to suburban shopping centres.

From around this time it is possible to see a number of related events – the closing of streets to traffic, the introduction of pedestrian streets, as well as a growing amount of research and publications promoting the concept of public spaces and public life. This 'Public Life oriented wave' in urban research and urban planning (Gehl, 1971, 2006) has now been around for some forty years, gaining momentum all the way along.

Europe's first major wave of pedestrian streets dating from the 1960s were primarily introduced to provide better conditions for customers in the commercial centre of the city. The streets were conceived as shopping streets, and pedestrianisation was seen as an urban response to the new suburban shopping malls that allowed people to shop without interference from traffic. Shopping bags dominated the city centres in the 1960s and 1970s, and the main activity was walking between the shops. At this first stage the streets truly were walking streets (Gehl, 1971, 2006).

The Danish capital, Copenhagen, serves as starting point and case study for this account concerning the dramatic changes in public life. For the past 40 years, researchers from the School of Architecture at the Royal Danish Academy of Fine Arts have systematically recorded the development and changes in the life and spaces of the city. Major studies have been conducted, in 1968, 1985, 1995 and 2005, making Copenhagen the first city in the world where the development of city life has been followed and documented over several decades (Gehl and Gemzøe, 1996; Gehl et al., 2006).

Case story: Copenhagen

The comprehensive surveys carried out in Copenhagen can be seen as part of this greater movement towards securing a renaissance for public space and public life in the culture of European cities (Gehl and Gemzøe, 1996). The Copenhagen studies were inspired by the introduction of Strøget, the main street of Copenhagen, as one of the first major pedestrian conversions in Europe. Strøget was closed to car traffic in November 1962 and by 1968 the new use of the street had stabilized and lent itself to investigations. What kind of life would be going on in such a traffic-free environment? And how were the other public spaces of the city used at this point? These were some simple research questions from the very first surveys, but the data thus gathered has made it possible to return regularly over the following forty years, in order to see how the spaces and the public life have developed. It is in this context that the dramatic changes in the character of public life have been recorded.

New and much more diversified activity patterns soon began to emerge towards the end of the 1960s and through the 1970s. The first outdoor cafés arrived, and the student revolution and 'flower power' movement brought people into the streets for political and cultural happenings. The trend was reinforced gradually as car parking was eliminated from the city squares and there was more room for city life. Soon city space was used for more political and cultural events, as well as quiet recreation and enjoyment. The number of cafés with outdoor service and the sheer number of café chairs in the city centre of Copenhagen has, during the four decades studied, almost exploded, from 3,000 in 1985 to 5,000 in 1995 and 7,000 in 2006. This development has been accompanied by an impressive extension of the outdoor season. Thirty years ago, the outdoor season in Copenhagen lasted two summer months, but it has now gradually been extended to cover almost the entire year. The café chairs now come out in early March and only go in after the closure of the equally new phenomena – the Christmas Markets.

In general, changes in the pattern of using city space have been far-reaching. A century ago, activities were almost exclusively necessary, forty years ago the primary focus was shopping, while recent decades have added a host of recreational activities and more cultural events, parades, happenings and exhibitions. Most recent is the wave of sport and exercise in public spaces.

The general feature of these changes is that, within the span of only a few decades, a work-oriented cityscape has become a city of leisure and enjoyment. Of course, the picture is not quite that simple, because working and shopping are obviously still going on, but now in parallel with urban recreation and many other pastimes. Another trend is that outdoor recreation and enjoyment are, as might be expected, predominantly a summer phenomenon, heavily dependent on the quality of city spaces as mentioned earlier, while winter city life continues to be more dominated by work-related activities and shopping. A distinct, two-season culture has evolved.

When people were interviewed in the 1970s and 1980s, their primary reason for being in the city centre was: 'shopping'. The same question asked in 2005 was more likely to generate the response: 'to be in the city'. By 2005 the city had definitely become a goal in itself, a destination in its own right. The growing trend of moving residences back to the city centres generally tells the same story (Gehl et al., 2006).

The new city life

Looking generally at the development of city life in Copenhagen over the past forty years, we see it undergoing a dramatic development after many years of languishing under pressure from car traffic. More people use the city and spend more time there, and we can see city life growing year by year. The days are longer as city life is expanded to include evening hours on days with good weather. The week is longer with more activities on Saturday and, most recently, Sunday as well. And, finally, as mentioned above, the outdoor season has been extended appreciably (Gehl et al., 2006).

The dramatic development in indirect digital communication has led to many predictions that public space will soon be replaced by cyberspace. The surveys from Copenhagen and many other cities, notably the recently published evidence of increase in city life in Melbourne (City of Melbourne and GEHL

1.4 2005. Strøget, now a mixed staying and walking street. Recreation and cultural activities play an increasingly greater role in the city scene.

PUBLIC SPACES FOR A CHANGING PUBLIC LIFE

Architects, 2004) certainly do not support this theory. During the years in which the new electronic options have burgeoned, city life has been markedly strengthened. The multiplicity of electronic pictures and messages would appear not to detract from public life but rather to provide extra inspiration to 'be present in person' and 'see things with your own eyes'.

From Canada in the west to Japan in the east, and from Australia in the south to Scandinavia in the north, street-side cafés are increasingly more numerous in the city scene. The users of city space are frequenting cafés by the thousands. A

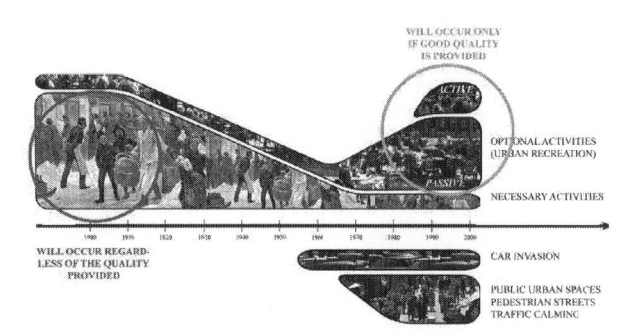

1.5 From necessary to optional activities. Development of public life from 1880 to 2005.

A graphic illustration of the dramatic changes in the character of city life during the twentieth century. Essential work-related activities dominate around 1900. The streets are crowded with people, most of whom have to use city space for their daily activities. The picture has changed appreciably by the year 2000. Essential activities play only a limited role because the exchange of goods, news and transport has moved indoors. In contrast, elective recreational activities have grown exponentially. Where the city once provided a framework almost exclusively for work-related daily life, the city hums with leisure- and consumer-related activities in 2000. Recreational activities set high standards for the quality of city space, and can be roughly divided into two categories: 1) passive staying activities such as stopping to watch city life from a stair step, a bench or a café, and 2) active, sporty activities such as jogging and skating. The timeline also shows when the car invasion hit Denmark in the mid 1950s. The pressure of car traffic and functional planning in the 1960s triggered a counter-reaction to reclaim city space. In the following forty years this reaction was reinforced, and developed nationally and internationally in an ongoing process.

hundred years ago, people went to the city because they had things to do and were forced to go, and while they were there they had many opportunities to look at life and meet fellow citizens. Today, where staying in the city is a choice, people need activities to keep them appropriately occupied for hours without attracting unwanted attention. Here, the cappuccino culture has acquired an important role as an excuse or rational explanation for the many lengthy stays in town. Both then and now, the meeting between people is a key city function, but people now use new formal explanations for spending time in the city. This is exactly where the cappuccino has come in handy.

Dramatic changes in living standards, working life and the economy have contributed in various ways to the new functions of city space over the past century. Households have shrunk. In the past hundred years, the size of the average household has been reduced from 4 to 1.8 persons. Young people study longer and start their families later than before. There are more single adults with their own dwellings and by now also more older residents in small households, due to increasing longevity. In many parts of the city, half of the dwellings have only one resident (Gehl *et al.*, 2006). People also have more room than they used to. Today there are more square metres per person in dwellings, at work, in shops and in businesses. A century ago, nine times more people lived in a typical urban quarter in Copenhagen than is the case in the same areas today (Gehl *et al.*, 2006).

On the job market, the relationship between work and free time has changed. If we look at the whole course of life, we have considerably more free time than we did a hundred or even fifty years ago. Staying in school longer means more years before work starts, and the marked increase in longevity means many more good years after retirement. And even when working life is at its most intense, longer holidays, weekends and days off allow more free time.

At the same time, production conditions and the economy have changed so that society on the whole has greater resources for consumption and pleasure. Changes in purchasing patterns and, for example, expenditure on holidays and travel attest to the new times.

Many other noteworthy changes in society, with varying impacts on city life, have taken place: the education of women

JAN GEHL

and their participation in the job market, children's long school days and institutionalized childcare, cars and increased mobility, and the increasingly more comprehensive indirect communication that allows pictures and contact to be exchanged via wireless networks, radio, TV, computers, the Internet, e-mails, video conferences and new interactive electronic systems. In various ways, all of these media provide the platform for changes in city life and the development of new city life.

New roles for public space and public life are redefined every day in a situation where daily life for many people continues to be steadily more privatized. The private sphere is growing as more and more functions are handled privately and individually: private dwellings, cars, offices, computers, washing machines, TV, shopping centres and other privatized solutions have taken the place of communal solutions to everyday problems.

Meeting other people is no longer an automatic part of daily life. While we have more resources, more time and more space, we are not necessarily one big happy family and, in fact, our direct meetings with other people can be few and far between. This paradox is precisely where we need to look when we seek an explanation while public life – now in new forms – is coming back to the public spaces in quantity.

City as meeting place in the twenty-first century

Throughout history, meeting other people has been the most important function and attraction of the city, and city space has had a central role as meeting place. In a changed society, city life in its various new forms can be considered to absorb and redefine the traditional meeting function in new ways. Now, as before, city space is the framework for people's meetings with society and each other.

In a society where concepts such as democracy, diversity and feelings of personal safety are considered important dimen-

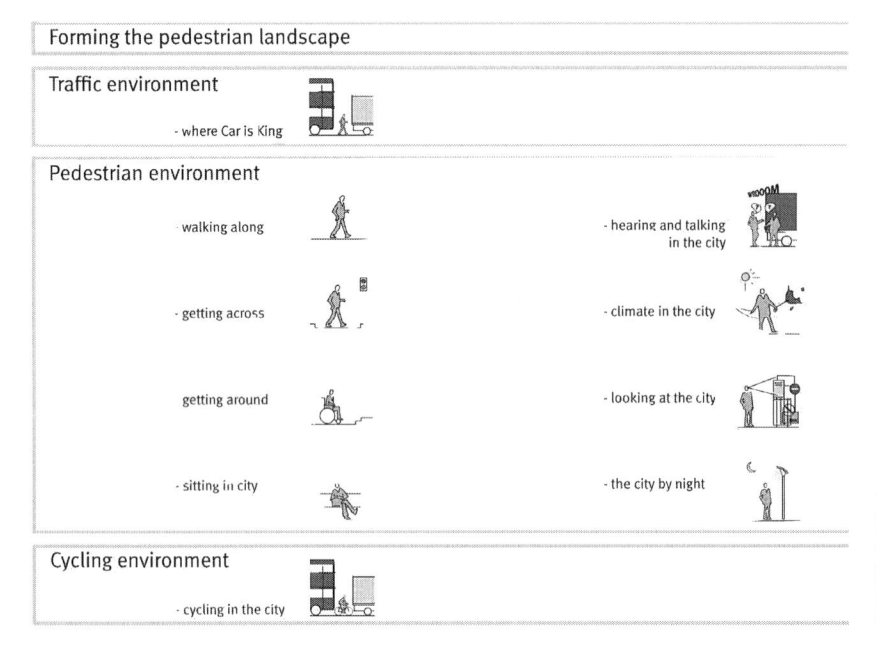

1.6 Quality of public spaces becomes increasingly important for the attractiveness of cities. Quality analysis of conditions for pedestrians (and bicyclists) in central London: overview of activity categories to be analysed (GEHL Architects, 2004).

sions, the extended use of public space must be seen as a very valuable development. And, for the same reason, modern requirements for good public space quality must naturally be honoured.

Now, as before, facilitating the meeting between people is the most important collective function of the city. The changing character of city life with its demands for good city space is a new expression of one of the most important functions of city culture: the meeting of people.

References

CIAM (1933) Athens Charter, 1933. CIAM – Congres Internationaux d'Architecture Moderne – founded in Switzerland in 1928, was an avant-garde association of architects, and a series of meetings (up to 1956), intended to advance both modernism and internationalism in architecture. The fourth CIAM Congress in 1933 (theme: 'The Functional City') consisted of an analysis of 34 cities and proposed solutions to urban problems. The conclusions were published as 'The Athens Charter' (so-called because the Congress was held on board the SS Patris en route from Marseilles to Athens).

City of Melbourne and GEHL Architects (2004) *Melbourne: Places for People*. Report to the City of Melbourne. Available at www.gehlarchitects.dk/images/melbourne_2004.pdf.

Gehl, J. (1971, 1980, 1987, 1996, 2006) *Life between Buildings*. Copenhagen: The Danish Architectural Press.

Gehl, J. and Gemzøe, L. (1996) *Public Spaces – Public Life, Copenhagen 1996*. Copenhagen: The Danish Architectural Press.

Gehl, J. and Gemzøe, L. (2001) *New City Spaces*. Copenhagen: The Danish Architectural Press.

Gehl, J., Gemzøe, L., Kirknæs, S. and Søndergaard, S. (2006) *New City Life*. Copenhagen: The Danish Architectural Press.

GEHL Architects (2004) *Towards a Fine City for People: Public Spaces and Public Life – London 2004*. Report to Transport for London and Central London Partnership. Available at www.gehlarchitects.dk/images/28780_tfl_public_spaces.pdf

Jacobs, J. (1961) *The Death and Life of Great American Cities*. New York: Random House.

'The health of the people is the highest law'
Public health, public policy and green space

Ken Worpole

CHAPTER 2

Introduction

The connection between public health and the provision of free, accessible, open green space – particularly in towns and cities – is obvious to most people. However, awareness of this connection has been muted if not entirely suppressed in terms of the public policy agenda for several decades. Tristram Hunt's history of Victorian municipal enterprise, *Building Jerusalem*, is only one of many accounts of British local government endeavour in the nineteenth and twentieth centuries to have described and celebrated the role which the proliferation of public parks was to play in the physical and spiritual renewal of the urban classes (Hunt, 2004). Indeed, a connection between local democracy and the ancient greensward was evinced in an editorial in the *Birmingham Daily Press*, cited by Hunt, which argued that, 'The self-government which is a peculiarity of the Anglo-Saxon race was brought out of their old German forests and planted here' (Hunt, 2004: 200).

For the Victorians the provision of parks was, therefore, symbolic of a wider commitment to the public good, and as much about character formation and citizenship as it was about physical well-being. Government interest in public health was also implicated in issues of national defence, including population control and fertility, although that is another story (Worpole, 2000). This broader perspective on the relationship between health and the provision of parks was later to change. The proliferation of new playing fields and recreation grounds witnessed after the First World War was predicated principally on concerns about the physical health of the masses – especially unemployed young working-class men. Social historians such as David Matless have subsequently found the motives of some of the grandees involved in various sports and fitness initiatives in this period decidedly mixed (if not militaristic), with very little to do either with notions of citizenship or democracy.

Yet the paternalism which informed both eras implied that local and national government bore a large degree of responsibility for the fitness and physical well-being of the population as a whole. This view has been in retreat for some time. It may well owe something to the reduced threat of war, or at least the assumption that future wars would be fought by technological means rather than large numbers of individual members of the

KEN WORPOLE

armed forces who needed to be physically fit. It also owes a lot to the pervasive ideology of neo-liberalism with its overweening ambition to bring about the demise of the 'nanny state'.

The periodisation of such political and cultural shifts and trends is an inexact science, but I would suggest that since the 1970s all the major political parties in the United Kingdom have espoused a 'less government is best government' approach to the health and leisure interests of citizens, leaving it as a matter of individual free choice how people wish to live their lives, look after their bodies and exercise their fertility. Individual physical well-being has moved from being a public health concern monitored and regulated by government to a market-place activity linked to individual wealth and lifestyle. Even such measures as mass inoculation to protect the 'herd immunity' of the population – as in the recent controversy over the combined MMR (measles, mumps and rubella) injection (Fitzpatrick, 2004) – have come to be seen as a matter of individual conscience, rather than as a matter of protecting the common good, by compulsion if required.

It therefore seems timely to look at some of the deeper political changes which have impacted on public policy in the twentieth century with regard to the provision, funding and management of green space. While in the second half of the last century this was largely a picture of a retreat from public health concerns in favour of individual choice over physical well-being and lifestyle, since the election of a New Labour government in 1997, the tide may just be turning. Arguments over the role which green and public space could and should play in social policy agendas have begun to be raised within government departments – even if largely behind the scenes. It could be that there is beginning to be a new political consensus emerging, in which public health, social policy and green space are seen to be in a vital and beneficial relationship to each other. What happened to bring this change about?

The 'tipping point' moment could be dated to 26 May 2004 when the UK House of Commons Health Select Committee 'Inquiry Into Obesity' stated that,

> Should the gloomier scenarios relating to obesity turn out to be true, the sight of amputees will become much more familiar in the streets of Britain. There will be many more blind people. There will be a huge demand for kidney dialysis. The positive trends in recent decades in combating heart disease, partly the consequence of the decline in smoking, will be reversed. Indeed, this will be the first generation where children die before their parents as a consequence of childhood obesity.
>
> (Select Committee, 2004)

The idea that without government intervention a large number of children might die of ill-health before their parents certainly stopped many policy-makers and politicians in their tracks. It is said that a politician thinks of the next election whereas a statesman thinks of the next generation. Here, clearly, was a matter of national concern with effects reaching far into the future, requiring a need for statesmanship and long-term planning after all.

Fortunately, from the mid-1990s onwards a series of independent surveys and government reports were published which not only charted the decline of public parks, but also analysed their prospects for the future, and the contribution they could make to public health concerns if revived. These included *Public Prospects* (Conway and Lambert, 1993); *Park Life* (Greenhalgh and Worpole, 1995); *People, Parks and Cities* (Department of Environment, 1996); *Towards an Urban Renaissance* (Urban Task Force, 1999); the Select Committee Report on *Town and Country Parks* (Department for Transport, Local Government and the Regions (DTLR), 1999); the Office of the Deputy Prime Minister (ODPM) report Planning Policy Guidance 17, *PPG 17: Planning for Open Space, Sport and Recreation* (ODPM, 2002a); the Urban Green Spaces Task Force Report, *Green Spaces, Better Places* (DTLR, 2002); the ODPM reports *Living Places: Cleaner, Safer, Greener* (ODPM, 2002b) and *Sustainable Communities: Building for the Future* (ODPM, 2003), among others. Many made a claim for putting well-managed parks back at the centre of urban life and leisure, on the basis of their continued popularity and use, despite their depleted condition. The notion of 'cleaner, safer, greener' became something of a mantra for those government departments asked to develop policies for neighbourhood and community renewal (ODPM, 2002b). The point of this chapter is to present the arguments for spaces and places which are 'cleaner, safer, greener – and healthier, too'.

PUBLIC HEALTH, POLICY AND GREEN SPACE

2.1 'Health of the People'. Plaque on the Public Health Service Department in the Metropolitan Borough of Southwark, opened 1937. Photograph by Larraine and Ken Worpole.

Live out of doors as much as you can

Anybody sitting on the upper deck of a bus travelling from the Elephant and Castle down Walworth Road in south London may well notice at eye-level a plaque on a building on the east side of the street, proclaiming, 'THE HEALTH OF THE PEOPLE IS THE HIGHEST LAW'. This plaque was unveiled on 25 September 1937 at the opening of the new Public Health Service Department in the Metropolitan Borough of Southwark. It is an icon of its era and the ascendancy of social democratic politics in many parts of Britain and Europe at the time.

Close by, the Peckham Health Centre, designed by the engineer Sir Owen Williams, opened two years earlier, in 1935. Also known as the Pioneer Health Centre, this included a swimming pool, gymnasium, theatre, nursery, dance halls, a cafeteria and games rooms, as well as medical facilities, set in parkland. It quickly became famous throughout the world as the most fully developed approach to public health care. Between the Walworth Road building and the Peckham Health Centre you can still find the Brockwell Park Lido, set in the grounds of the magnificent Brockwell Park, the lung of this part of south London for over a hundred years. The lido opened in 1935, one of more than twenty other lidos built in public parks in London in the same decade, and designed by Harry Arnold Rowbothan and T.L. Smithson, both of whom worked in the London County Council Parks Department for much of their lives (Worpole, 2000).

This period was an era of great political investment in public health, a pattern common across much of Europe. Health centres, clinics, recreation grounds, sports fields, lidos and nursery schools were among the most innovative new building types being developed, and modernist ideas and ideals in architecture were rapidly transplanted from one country to another. One of the key tenets of the modernist movement was the

KEN WORPOLE

integration of the building and its landscape setting, overcoming the distinction between indoors and outdoors. Berthold Lubetkin's Finsbury Health Centre in north London, completed in 1938, was partly based on designs Lubetkin had originally drawn for a Palace of Soviets. On the large curving walls of the foyer, the designer Gordon Cullen had been commissioned to produce two murals, based on the slogans, 'LIVE OUT OF DOORS AS MUCH AS YOU CAN' and 'FRESH AIR NIGHT AND DAY'. The murals have gone but the building itself still functions successfully, otherwise unchanged, and is immediately adjacent to Spa Green Fields park, beautifully re-designed and renovated in 2006.

Many today would question the quasi-anthropological language then used to describe these progressive experiments in public health provision, and the technical prospectuses were often couched in the vocabulary of positivistic social engineering. A close look at the documentation which surrounded their planning, design and monitoring of effects suggests more than a hint of social and genetic determinism, of apprehensions about 'fitness to marry and breed' or of declining class vitalism. Nevertheless they represent a period in which progressive politics, a concern for public health and architecture marched in step, and in the same direction.

This concern with the relationship between modernism and health was not just about buildings. The British landscape architect, Sir Geoffrey Jellicoe, wrote in the 1970s that, 'It is only in this present century that the collective landscape has emerged as a social necessity' (G. and S. Jellicoe, 1975). For the whole culture of town planning in the twentieth century was about establishing the right balance between the built environment and the proper allocation of green space for leisure and recreation.

The architecture and landscape design inspired by the political aspirations of early twentieth-century social democracy remains a legacy worth celebrating – because it has lessons for us today. It was precisely this history I sought to recover in *Here Comes the Sun* (Worpole, 2000), essentially a story about how, at the beginning of the twentieth century, social reformers, planners and architects tried to re-make the city in the image of a sun-lit, ordered, healthy utopia. The astonishing growth in demand for new institutions and landscapes in Europe in the early years of this century arose directly from the rise of democracy, and a newly enfranchised citizenry and its political organizations confidently demanded better housing, health, education, transport, public landscapes and even leisure facilities – and a number of architects and landscape designers of the modern school energetically responded.

However, this should not be understood as a top-down movement only. The initial passion for popular fitness and exercise came from below. From the 1890s onwards, leisure activities, particularly walking, cycling, camping and trips to the countryside and seaside, were associated with political and health reform, and came from grassroots movements and new forms of associational life and culture. In Germany this culture was called *lebensreform*; in Britain it was often referred to as the 'art of right living'. The invention and rise in popularity of the bicycle was strongly associated with women's emancipation (and the rational dress movement). The growth of rambling clubs was part of the culture of nonconformism, temperance and the proselytising activities of the early socialist movement. William Morris's creed that, 'Fellowship is heaven, and lack of fellowship is hell; fellowship is life, and lack of fellowship is death' (Morris, 1968: 51), inspired this link between leisure and a vision of a different, and better, society. The Clarion Cycling Club formed in 1894 was socialist in origin; by 1924 there were 24 rambling clubs in Sheffield alone, nearly all associated with the Clarion movement. This pattern was repeated across the country. Access to the countryside and the right to roam were key demands of this burgeoning life-reform culture, culminating in the famous Kinder Scout mass trespass of 24 April 1932, which eventually led to the 1939 Access to Mountains Act (Taylor, 1997).

This politics of the outdoors was an international phenomenon in the early twentieth century, although by the 1930s it was beginning to bifurcate into distinct left-wing and right-wing attachments and organizations, particularly in Europe. Harvey Taylor dismisses, however, any connection between the British open-air fraternity and that of the German *wandervogel* movement: 'The peculiarly British outdoor movement was rooted in the language of open-air fellowship and the rights of the freeborn Englishman, or the Scottish stravaiging tradition of roaming at will, rather than atavistic romanticism' (Taylor, 1997: 4).

In Britain, participation in the outdoor life movement reached to the very top. In the last year of the post-war Labour government, a government minister, Lewis Silkin, introduced

PUBLIC HEALTH, PUBLIC POLICY AND GREEN SPACE

the 1949 National Parks Bill with the words: 'This is not just a Bill. It is a people's charter – a people's charter for the open air' (Matless, 1998: 248). Like Barbara Castle, Hugh Dalton, Chuter Ede and other cabinet members who were active members of the Ramblers' Association, Silkin believed that outdoor recreation was the key to the future of public health, as well as to a more equitable and democratic society. Yet less than twenty years later, the culture of public health had changed irrevocably, no longer regarded as the principal means to a better society, and a matter of national concern, but left as a matter of individual choice.

The decline in government support for outdoor facilities was symbolised in the late 1960s with the closure of many outdoor swimming pools, lidos and children's paddling ponds. The decline of spending on public parks since the 1970s is even more well documented. The 'Public Park Assessment' undertaken by the Urban Parks Forum (now GreenSpace) revealed that the United Kingdom's 27,000 parks suffered an estimated £1.3 billion cuts over the previous two decades (Urban Parks Forum, 2001). Years of local authority budget-trimming – often exacerbated by the introduction of Compulsory Competitive Tendering – had seen a depletion of staff (particularly skilled staff), and the neglect or removal of many historic park features. The Assessment estimated that the percentage of original features lost to Britain's parks were: park lodges (24%), glasshouses (35%), fountains (40%), mansions (42%), bandstands (50%), monuments (55%) and paddling pools (60%). The once-loved town park was in danger of becoming a boarded-up war zone, while distinct varieties of parks, each with its own design history and typology, were being turned into anonymous green deserts, as a result of 'one-size-fits-all' management and maintenance regimes.

2.2 Maryon Park Café. 'The once loved town park was in danger of becoming a boarded-up war zone.' Photograph by Larraine and Ken Worpole.

There were, it is true, exceptions to this pattern of decline, especially with the creation of regional and country parks, but within most towns and cities of Britain the public realm – as a series of spaces and places, well managed and looked after – was in retreat. Not only were green spaces poorly looked after, but distinctive management and maintenance regimes appropriate to different types of green space were being homogenized in the name of economies of scale. Historic cemeteries were being levelled and crudely machine-mowed; ornamental gardens were stripped of their horticultural variety; allotments were neglected and in some places sold for development; majestic park trees were replaced by smaller-scale varieties ('lollipop trees'). A variety of park types was in many places succeeded by a collection of green deserts.

There are a number of reasons for the loss of political interest in outdoor recreation. They would certainly include:

- the rise of the private car and its dominance of streets, public spaces and modes of local travel;
- the growing commercialization of fitness facilities and sports activities (as part of a global branding exercise for sports-based consumer products);
- the role that international competitive sports are playing in promoting national identities on a global stage (leading to a concentration of public resources on elite sports);
- the rise of the indoor leisure centre as an icon building of social regeneration and urban renewal (areas of high unemployment in Britain in the 1970s were quickly littered with sports halls and leisure sheds);
- the influx into local authority leisure departments of a new generation of graduates with sports and leisure administration degrees (who favoured indoor sports and arts facilities over parks).

The results of these trends were evident in a graph illustrating the report, *Green Spaces, Better Places*, published in 2002 by the Urban Green Spaces Task Force (DTLR, 2002). It revealed that spending on 'Urban parks and open spaces' had dropped from 44% to 31% of local authority spending on leisure between 1976/77 and 1998/99. In the same period there had been, it is true, an increase in spending on 'Country parks, nature reserves and tourism', from 7% to 17%, but the tourism budgets of local authorities were usually dedicated to promoting visitor spending on hotels, heritage sites and commercial leisure attractions.

The fact is that the proportion of local authority budgets spent on the 'vernacular outdoors', defined as parks, footpaths, playgrounds, cemeteries, cycle paths and pedestrian networks, declined throughout the 1970s, 1980s and 1990s, year on year. In the same period, leisure was increasingly defined as a form of cultural consumption, to the extent that in 1995 the National Playing Fields Association claimed that the Department of Heritage spent three pence on the needs of children for every £100 spent on adult leisure (Wallace, 1995: 8).

These trends caused members of the Urban Green Spaces Task Force – set up by the government to look into the crisis facing Britain's parks – to warn in 2002 that there were now 'two cultures of leisure' operating within urban communities: one of indoor leisure centres and private fitness clubs, and the other of urban parks and other outdoor recreational spaces. Yet, contrary to all government policies ostensibly geared to tackling social exclusion, evidence suggested that public spending on active leisure was increasingly being used to support the lifestyles of car-driving, fully employed and mostly male professionals, who used indoor leisure centres, rather than the much larger group of people who used parks.

According to leisure analysts Graham Jones and Paul Greatorex, a survey of 155 sports halls and swimming pools in England revealed that, 'The use of sports halls is dominated by the non-manual socio-economic groups; professional and managerial classes also tend to have greater representation in the use of swimming pools. The use of sports halls and swimming pools is dominated by those working full-time' (Jones and Greatorex, 2002). Somewhat perversely, fitness fanatics also felt obliged to drive to their gruelling encounter with a walking or cycling machine, as was made evident in a survey of 1,000 indoor leisure centre users in England in 2001, which noted that 'a staggering 89% of customers travel by car' (Hill, 2002: 24). By contrast, surveys of park users show that 70% of them walked to the park.

Not only was public spending going in precisely the opposite direction to that implied in New Labour's 1997 commitment to tackling social exclusion, by directing public resources to the 'haves' rather than the 'have-nots', it was also palpably failing to deliver value for money. The Task Force noted that in 2001,

2.3 Cycling in Vondel Park, Amsterdam. In the Netherlands, 50% of people claim to walk or cycle regularly.'
Photograph by Larraine and Ken Worpole.

2.4 Aerobics in Stockholm park. Free outdoor exercise sessions are a common feature in Stockholm's parks in the summer.
Photograph by Larraine and Ken Worpole.

KEN WORPOLE

public spending on parks was estimated to be around £600 million for the year, but this achieved 2.5 billion visits. On the other hand the £400 million spent in the same year on indoor leisure facilities only achieved 100 million visits. In terms of conventional 'value for money', public investment in outdoor recreation might achieve better returns with regard to public health than an over-concentration on indoor leisure.

One of the reasons is that indoor sports have something outdoor recreation lacks: a public body dedicated to promoting their benefits and use. What is more, indoor and organized sports not only have a professional agency, Sport England, to argue for increased funding, they also enjoyed their own dedicated Sports Lottery Board – as did the Arts. Until the establishment of the Commission for Architecture and the Built Environment's unit, CABE Space, in 2003, Britain's parks and open spaces had no dedicated minister, agency or lottery board (though the Heritage Lottery Fund's 'Public Parks Initiative', established in 1996, has achieved wonders in supporting parks) to fight their corner, despite the phenomenal popularity of park use. As Jones and Greatorex pointed out, 'many very popular activities are disenfranchised because they do not have a strong administrative structure [since] the vast majority of what might be termed physical recreational activity takes place away from conventional facilities' (Jones and Greatorex, 2002).

The Urban Green Spaces Task Force report of 2002 recommended a modest £500 million over the next five years to halt the decline in parks (a modest sum compared to the £10 billion which Sport England estimated was needed to bring indoor provision into the twenty-first century). Some of this new funding is beginning to flow, and many parks are now improving. Yet the bigger issue of how best to promote public health through informal recreation goes unresolved.

There is enough evidence now to convince politicians. The British Heart Foundation has claimed that a third of under-7 year olds fail to reach the minimum activity levels recommended by the National Health Service – and by the age of 15, two-thirds of girls are classified as inactive. Seven out of ten school leavers abandon formal kinds of physical activity. Two-thirds of 9–11 year olds in the United Kingdom are dissatisfied with the quality of outdoor play facilities where they live. For 15–16 year olds this rose to 81%, higher than any other European country (Worpole, 2003). In the Netherlands, 50% of people claim to walk or cycle regularly, compared with under 10% in the United Kingdom. England's overall participation rate in sport and physical activity is 21%, compared with 52% in Finland, 45% in Australia and 38% in Canada (Culf, 2005). Obesity in England has grown almost 400% in 25 years, with three-quarters of the adult population now overweight, with some 22% declared obese (Select Committee, 2004).

Why, given this escalating crisis of public health, are public policies not more focused on supporting popular forms of recreation outdoors? It is not just a problem created by national government, as it has been local authorities who have principally overseen the transfer of funding from outdoor to indoor provision. It is still the case that many elected local councillors regard the creation of leisure centres as being more 'modern' than supporting traditional parks, as well as providing better PR opportunities – without apparently doing even the most elementary cost–benefit analysis. New Labour's attitude to public health and life outdoors, too, has changed radically over the past fifty years. When government ministers conjure up a picture of outdoor space in twenty-first-century urban Britain, it is often peopled with feral youths, burnt-out cars, graffiti-covered walls and upturned park benches. 'Safer indoors', you can almost hear them sigh. The dream of the fellowship of the great outdoors has fled. Instead it has become a 'degraded realm' in the upper reaches of New Labour's political imagination, and this ambivalence about the value of outdoor life and the pleasures of the public domain has permeated public consciousness everywhere in Britain.

The favouring of formal indoor leisure goes against the grain of other government policies – on transport, crime, urban regeneration and community-building – which seek to encourage walking and cycling as the best way to animate streets and bring about greater security through 'natural surveillance'. So, as some government departments began to develop policies to strengthen the relationship between having 'cleaner, safer, greener' neighbourhoods and communities as a way of framing new forms of social and public health policy, other departments turned in the opposite direction.

Ironically this largely resulted from the, perhaps unexpected, winning of the 2012 Olympics and Paralympics bid for London. Until this occurred, national sports policy was generally moving towards greater support for popular participation in sports and recreation, if mostly of a formal kind. This was partly to counteract the stagnation in popular participation in the 1990s (after increases in the 1980s) (Coalter, 2004). Despite claims by government ministers that winning the bid for hosting the Olympics provided an incentive for all of the population to get involved in sports, research commissioned after the 2002 Manchester Commonwealth Games showed that the Games had no measurable impact at all upon public participation – if anything membership of sports clubs in Greater Manchester declined (Coalter, 2004). Even Sport England itself admitted that, 'Hosting events is not an effective, value-for-money method of achieving a sustained increase in participation' (Conn, 2006: 6).

Furthermore it is likely that funds (including lottery monies) will be diverted away from other good causes, including outdoor recreation, towards supporting Olympic athletes. A headline in the *Scotsman* newspaper on 9 January 2004 proclaimed that the 'London bid may cost sport in Scotland £40m'. An editorial in the *Observer* newspaper on 26 March 2006 noted that, 'In the next few years, £340 million is to be taken away from the Sports Lottery Fund to feed the Olympics project, with the possibility of a further £410 million being diverted through re-arrangement of percentage shares of National Lottery proceeds' (*Observer*, 2006: 24). These transfers of funds suggest that community sport and recreation will suffer from an over-concentration on achieving a successful Olympics.

While it is not likely that sports policy will promote the recreational and health benefits of investment in parks and green spaces – nor remind politicians of its value-for-money record – the communitarian argument which still holds sway at the Department of Communities and Local Government (formerly The Office of the Deputy Prime Minister) is likely to be more effective. The 'new localism' – as current concerns with the neighbourhood are now termed – seeks to bring decision-making and control over the quality of neighbourhoods as close to the residents and other stake-holders as possible. It is partly inspired by a number of public opinion surveys which put a desire for safer and better-maintained environments high on people's list of local priorities, as was evident in *Living Places: Cleaner, Safer, Greener* (ODPM, 2002b).

According to the Deputy Prime Minister in this report, 'Successful, thriving and prosperous communities are characterised by streets, parks and open spaces that are clean, safe, attractive – areas that local people are proud of and want to spend their time [sic]' (ODPM, 2002b: 5). It is also suggested that 'Public spaces mean everywhere between the places where we live and work' (ODPM, 2002b: 9). This seems like a good basis for promoting the value of green space.

Conclusion

The role that informal outdoor recreation has played in public health has been too little acknowledged in public policy in recent decades. This has coincided with the loss of status of parks and public spaces, and other recreational networks. Which is cause and which effect may be debated. Issues of personal fitness and well-being have been increasingly left as a matter for individual choice, to be provided for in the commercial market-place. Spending on public health, through the National Health Service, almost exclusively takes the form of spending on the treatment of poor health, including treating the effects of sedentary lifestyles, inappropriate diet, as well as the mental health aspects of isolation and physical self-neglect. Too little is spent on preventative health measures. If only a minute fraction of what is spent on the NHS were diverted to the improvement of parks and the wider public domain, then there might genuinely be an urban renaissance. Meanwhile notions of sporting excellence and the international competitiveness of sporting activity – as a result of commercial sponsorship and global prestige – have been espoused by politicians, often at the expense of popular forms of informal physical recreation. From time to time drug scandals in professional sports remind us that the moral high ground once occupied by the competitive ethos is being undermined from within.

In the United Kingdom, the principal policy driver for improving opportunities for outdoor recreation is currently the 'greener, safer, cleaner' approach for neighbourhood renewal, advocated by the Department of Communities and Local Government. This has the potential to harness local and national resources in creating a high quality network of streets, parks, pedestrian and cycle routes, which in turn could also provide real benefits for transport and environmental policy too.

2.5 Mile End park. The Mile End Park in east London, created in the 1990s, offers well designed pedestrian and cycling links from one part of the city to another. Photograph: Larraine and Ken Worpole.

There is much in recent years to give guarded cause for optimism, arising from the renewed national and local government interest in parks and green spaces as well as in the work of organizations such as Groundwork and the British Trust for Conservation Volunteers – in the success of the Green Flag Awards Scheme, in the funding programmes of the Heritage Lottery Fund's 'Parks for People' programme, in the advocacy work of GreenSpace and CABE Space, in the network capacity-building of organizations such as Sustrans and many local and regional walking and cycling campaigns.

Yet the case for greater public investment in parks and green space networks is not simply about responding to current concerns with childhood obesity, the sedentary lifestyles of those living in the age of the Internet, or the dominance of the car in transport policy. It is also about creating a sense of attachment to place and to other people, through the greater democracy and human engagement of life outdoors. The park, like the street or the seaside beach has, historically, been an astonishing arena for forms of conviviality and collective pleasure. While parks and public spaces can also be danger zones at times, they are more likely to act as places where the rules of public life and citizenship are tested and formed. In this sense they are not just about improving the physical health and well-being of people as they go about their daily lives, but about creating more reciprocal forms of social life as well. There is no sustainable future without them.

References

Coalter, F. (2004) 'Stuck in the Blocks', in A. Vigor, M. Mean and C. Tims (eds) *After the Gold Rush: A sustainable Olympics for London.* London: Demos & IPPR.

Conn, D. (2006) 'Laughing at the Finnish is a non-starter for London', *The Guardian Sports Section*, 26 July 2006, p. 6.

Conway, H. and Lambert, D. (1993) *Public Prospects: Historic Parks Under Threat.* London: The Garden History Society and The Victorian Society.

Culf, A. (2005) 'Fitness campaign goes to extremes', *The Guardian*, 30 August 2005.

Department of Environment (1996) *People, Parks and Cities*, London: HMSO.

Department for Transport, Local Government and the Regions (DTLR) (1999) *Town and Country Parks Report.* London: HMSO.

Department for Transport, Local Government and the Regions (2000) *Towards an Urban Renaissance.* London: DTLR/Spon Press.

Department for Transport, Local Government and the Regions (2002) *Green Spaces, Better Places.* London: DTLR.

Fitzpatrick, M. (2004) *MMR and Autism.* London: Routledge.

Greenhalgh, L. and Worpole, K. (1995) *Park Life: Urban Parks and Social Renewal.* Stroud: Comedia & Demos.

Hill, M. (2002) 'Internet Customer Surveys', *Leisure Manager*, February 2002. Reading: Institute for Sports, Parks and Leisure (ISPAL).

Hunt, T. (2004) *Building Jerusalem: The Rise and Fall of the Victorian City.* London: Weidenfeld & Nicolson.

Jellicoe, G. and Jellicoe, S. (1975) *The Landscape of Man.* London: Thames & Hudson.

Jones, G. and Greatorex, P. (2002) 'The pool, the hall, the pitch', *Leisure Manager*, April 2002. Reading: Institute for Sports, Parks and Leisure (ISPAL).

Matless, D. (1998) *Landscape and Englishness*. London: Reaktion Books.

Morris, W. (1968) 'A Dream of John Ball', in *Three Works by William Morris*. London: Lawrence & Wishart, p. 51.

Observer (2006) 'Don't make grassroots pay for Olympics', 26 March 2006, p. 24.

Office of the Deputy Prime Minister (ODPM) (2002a) *Planning Policy Guidance 17: Planning for Open Space, Sport and Recreation*. London: ODPM.

Office of the Deputy Prime Minister (2002b) *Living Places: Cleaner, Safer, Greener*. London: ODPM.

Office of the Deputy Prime Minister (2003) *Sustainable Communities: Building for the Future*. London: ODPM.

Select Committee (2004) *House of Commons Select Committee on Health Third Report*. Available at: www.publications.parliament.uk/pa/cm200304/cmselect/cmhealth/23/2303.htm.

Taylor, H. (1997) *A Claim on the Countryside: A History of the British Outdoor Movement*. Edinburgh: Keele University Press.

Urban Parks Forum (2001) *Public Park Assessment*. Reading: Urban Parks Forum.

Urban Task Force (1999) *Towards an Urban Renaissance: Final report of the Urban Task Force chaired by Lord Rogers of Riverside*. London: Department of the Environment, Transport and the Regions.

Wallace, W. (1995) 'Grounds for Complaint', in *Nursery World*, 8 June.

Worpole, K. (2000) *Here Comes the Sun: Architecture and Public Space in 20th Century European Culture*. London: Reaktion Books.

Worpole, K. (2003) *No particular place to go? Children, young people and public space*. Birmingham: Groundwork.

# Playful nature

What makes the difference between some people going outside and others not?

Catharine Ward Thompson

CHAPTER 3

Introduction

There is an expanding interest in people's use of open space near the places where they live and work, reflecting a number of different themes and concerns in current British government policy and planning – social inclusion, environmental justice, accessibility and healthy lifestyles (Land Use Consultants, 2004; CABE Space, 2004). These are reflected in initiatives such as The Department for Communities and Local Government's 'cleaner, safer, greener' campaign and the Sustainable Communities Plan (DCLG, 2003) and effected through public health programmes such as 'Paths to Health' in Scotland and 'Walking the Way to Health' in England, initiatives supported by the British Heart Foundation, Scottish Natural Heritage and the Countryside Agency. There has been a growing body of work looking at primary school aged children and their access to outdoor environments, particularly environments where natural elements predominate (Faber Taylor and Kuo, 2006). Until recently there has been far less work on teenagers and outdoor environments (Travlou, 2006; Ward Thompson et al., 2006) and on older people and their access to the landscape (Sugiyama and Ward Thompson, 2005), and very little indeed on intergenerational issues in relation to outdoor space and place, although this too is now recognized as important.

This chapter draws on recent research to reflect on how people engage with their environments and what makes the difference between some people going outside and others not. Going outside can mean simply going outside one's home, going anywhere that is open to the sky, from back gardens and courtyards to urban streets and parks, as well as to more remote countryside and coastal areas. For most people, day-to-day or regular outdoor use is likely to be relatively local to home, work or school. This chapter explores what matters to children and teenagers in this context, and what matters to older people. What might make a difference in people's lives and where are the challenges for planners, designers and managers of open space? Since people's enjoyment of outdoor places, enjoyment *in* outdoor places, is clearly one crucial factor, we might ask how natural environments can be pleasurable ones. How can we provide the best environments to support the playful natures in all of us, young and old?

CATHARINE WARD THOMPSON

The importance of play for children's development is well recognized (Cole-Hamilton et al., 2001; Richer, 2005) and, for many adults, such messages about the value of play will be reinforced by personal memories of those places rich in experience for them as children, often places they could go to on their own or without adult supervision (Hart, 1979; Moore, 1986; Lohr et al., 2000; Kyttä, 2004). Research suggests that there is an important link between play and natural environments (Hansen, 1998; Thompson, 2005; Faber Taylor and Kuo, 2006) and that there is an element in such experience which becomes part of a person's nature, from childhood onwards. Positive memories of unstructured play in woodlands and natural spaces in childhood are associated with looking to natural environments for their therapeutic or recreational value in adulthood (Bingley and Milligan, 2004; Everard et al., 2004). We might therefore ask what kind of access children and young people have to natural environments today, and what they think about their own experience as they mature. We also need to understand the perspective of adults and older people. What might make the difference for a middle-aged person or someone in their late seventies in terms of going out or staying in, for example? How important are physical aspects of the environment and what kind of difference might we make through our plans and designs?

Subsequent chapters in this book will address detailed aspects of access to and engagement with natural and outdoor environments, for different ages and socio-cultural groups. The aim here is a broader overview of the issues, drawing on recent research carried out in different parts of Britain by OPENspace Research Centre, as part of our remit to explore issues relating to inclusive access to outdoor environments.

Natural environments

Our work has added to the understanding of what people of various ages think about parks, woodlands and 'natural' places they might go to, and what they like to do there (see Box 3.1).

In a survey about woodland use by urban communities in central Scotland undertaken for the Forestry Commission (Ward Thompson et al. 2004), people were clear that they like woodlands to be free from rubbish, natural in appearance and readily

---

Box 3.1 What people like about visiting nearby woodlands: quotations from urban communities in central Scotland

'It's the noise I like. I like the trees, the rustling and the smells, and the water – the burn's on the walkway.'

'Out walking one winter years ago . . . We had had a really heavy fall of snow down on the walkway and the trees were all really heavy, and it was just like a tunnel with all the trees, and there was a stag down on the walkway. I remember that. It was so beautiful, so quiet.'
Women from mother-and-toddlers group

'You can just go away by yourself. You can just disappear and nobody can see you . . . you can't do that in the city, you can't just keep walking, walking, walking.'

'I find it's quiet, it gets you away from everyday life, basically. You just go away and be in a world of your own sometimes. You can go away if you're angry at anything, just go away and get yourself all calmed down.'
Unemployed men and women

'I like the bit up the wood, by the quarry. You can sit up at the top . . . and see the whole of Edinburgh.'
Teenager

---

accessible. Frequent (daily or weekly) visitors were more likely to use woodlands within walking distance of home, while less frequent visitors cared more about going to woodlands with good signage and information boards, rather than visiting places close to home. The most popular activities were seeing wildlife, going for family walks and taking children to play; getting fresh air and walking the dog were also popular. Men are more likely than women to visit woodlands on a daily or weekly basis, and more likely to go walking on their own, although being with a dog is also significantly correlated with walking alone. Perhaps surprisingly, given much of the previous research (e.g. Burgess, 1995) and publicity surrounding fears and concerns for safety in such places, the survey showed a strong sense, across all age groups, that people felt at peace in woodlands; people over 25 were also very likely to feel at home in woodlands. Indeed, most people strongly disagreed with the suggestion that they might consider woodlands scary or feel vulnerable and fear having an accident in woodlands, although women and older people were less dismissive of these concerns than men.

Sixty-six per cent of respondents in the central Scotland survey visited woodlands at least once a month. The findings reinforce the value of community and urban woodlands as familiar places where people can have a peaceful experience of nature. The choice of woodlands for recreational purposes appears to be driven mainly by proximity to people's homes, although signs and information and a tidy and welcoming appearance were factors particularly important for infrequent woodland users. Most people would ideally like a compromise between a very wild and natural appearance in their local woodland and the facilities and management associated with a more formally designated woodland or country park.

Despite the generally positive attitudes to woodlands as places where people feel at home, the aspects most constraining to use of the woodlands are issues of safety and environmental abuse. Although people were generally not concerned for their own safety, they often mentioned fears about safety in the context of concern for others, especially children, and older people were more likely to fear the consequences of injury or accident in a remote place away from help. Such fears are principally constraints on people wanting to visit woodlands alone and, although many people mentioned the desirability of having a ranger or warden on site, most such people would not visit woodlands unless they were accompanied by a friend or relation, in any case. Dumping of rubbish and general littering of forests and woodlands does not always deter people from visiting, but it does detract from their experience.

In another study based on the East Midlands, a region of England with a diversity of urban and rural communities, OPENspace surveyed visitors to 'natural' sites in the region, ranging from urban parks to wild and remote countryside areas, on behalf of English Nature (Bell et al., 2004). Even in a region whose towns contain a notably diverse urban population, there was under-representation among site visitors from minority ethnic and black communities (3.3% of visitors compared with 6.6% of the East Midlands population), people with disabilities (9.6% of visitors compared with an East Midlands average of 20%), adults under 24 (4.5% of visitors compared with an East Midlands average of 7.8%), people over 64 (15.0% of visitors compared with an East Midlands average of 21.4%) and women (44.4% of visitors compared with an East Midlands average of 50.88%). The proportion of women visiting alone was half that of men (7.88% vs 15.4%). Even allowing for the overlap of categories, this reflects other research on access to the countryside (e.g. Countryside Agency, 2004), which demonstrates that a majority of the population is actually under-represented in visits to natural or green areas and, by inference, that a minority of the population (white, young to middle-aged, able-bodied and male) is still dominant as users.

The East Midlands study confirmed that what visitors like most about the physical site qualities of the green spaces they visit are naturalness in appearance and freedom from rubbish. Signs and information were again important but to a lesser degree. Walking the dog, exercise and leisure were the most likely reasons for visits to green spaces but people's attitudes to activities show that they care most about relaxing, children learning about nature, and enjoying wildlife. People were very clear that green spaces were places where they felt peaceful and where they had a sense of freedom and affiliation with nature.

It is therefore not surprising that East Midlands survey respondents felt very strongly that green spaces are important for communities. In general, they did not feel uncomfortable or vulnerable in the kind of green spaces they visited, although, again, men expressed these views more strongly than women. Young adults under 25 years old were least positive about feeling an affiliation with nature and more likely to feel uncomfortable in green spaces. Their feelings of vulnerability matched those of the over-55s, being greater than the age groups in between.

Focus group discussions held with people from minority ethnic groups in Nottingham and Leicester, towns in the East Midlands with diverse ethnic populations, revealed a more particular perspective and highlighted some of the challenges for these groups in accessing green spaces and engaging with natural environments. As with other respondents, the proximity of green spaces is a key factor and local spaces are much more likely to be visited than sites at a distance. But participants felt there was a lack of information targeted at their communities and a lack of creativity in engaging young people from minority ethnic groups (see Box 3.2). There was a desire for guidance from knowledgeable people about how to introduce different communities to the natural environment and a need for innovative ways to facilitate access to information that would be meaningful to different groups.

The findings from these two studies in rather different parts of Britain reinforce each other in many ways. Woodlands and other green spaces close to people's homes are vitally important in making it easy for people to get outdoors and have contact with the natural environment. For planners, designers and managers, another clear message is that a natural and diverse appearance and freedom from rubbish are important to people, many of whom feel an affiliation with nature and appreciate green spaces for offering both a sense of peace and opportunities for children to play and to learn. The presence of signs, information boards (and other kinds of information) and evidence of site maintenance can be predictors of whether or not people will visit an outdoor site such as a woodland. Such physical signs of care and good management appear to be important in signalling a sense of welcome and evidently make a difference to people who are not regular visitors at present (see Boxes 3.1 and 3.2). But what comes out far more strongly than these physical attributes, as a predictor of how often people visit natural or green places, is their remembered childhood experience.

Box 3.2 Challenges in accessing green spaces and natural environments for minority ethnic groups: quotations from minority ethnic communities in Nottingham and Leicester, East Midlands

'There are some of them [local youth] interested, but our main problem is how to get there. You know they live in the local area and they just look for the nearest park, they can walk down. You tell them to go somewhere about three, four, five miles away and they can't.'

'I run a voluntary girls group, basically because girls don't have anything, anywhere socially, and it's only recently that we've started thinking about doing something environmentally friendly with them but we're not trained . . . I don't think that way myself and I don't know how to be creative about it to get them involved . . . Our ages range from the age of ten up to any age, because we cater for whoever wants to come.'

'If we know a lot more then we can motivate them [Asian young people]. You know when organizations have these open days . . . all the usual activities are there, like the henna, and it gets a bit boring, nothing out of the ordinary.'

'It's net-working and partnership working, there are bodies that are looking for information but they don't know the people who have got the information.'

'I've got maps, detailed maps, of nearly every park in Leicestershire and nobody knows how to get them off me and I don't know how to get them to other people. We need some sort of central body, like a local directory of what's on.'

The connection with childhood

Simple observation of young children's behaviour in most outdoor places will remind us of the importance of engagement with natural elements. Watching a two-year-old who has found a puddle to splash in, or a stick to poke in the earth, illustrates how much interest there is in elements of the environment that are changing and responsive to physical intervention (see Boxes 3.3 and 3.4). As children get older, their engagement with and manipulation of the natural environment becomes more sophisticated; experiences such as making secret places and building dens appear to be an important part of growing up (Sobel, 1993; Kjørholt, 2003; Bingley and Milligan, 2004; Barnard, 2006).

Our own research strongly reinforces the importance of childhood engagement with nature. People who remember frequent and regular visits to natural outdoor environments as children are more likely to go to such places as adults and will often have strikingly different responses to such environments compared with people whose childhood experience was limited in this

3.1 A sign that welcomes newcomers to woodlands and countryside in Clackmannanshire, Scotland.

Photograph by Katherine Southwell, © OPENspace.

3.2 Well-managed woodland entrance, Gartmorn, Scotland.

Photograph by Katherine Southwell, © OPENspace.

respect. While we have to allow for the possibility that memories of childhood may not be accurate, or may be biased in favour of subsequent adult activities and preferences, we certainly found evidence of a significant relationship between people's reported childhood experience of nature and their adult engagement with the natural environment. There is rich evidence of vivid childhood memories of the outdoors for many people and evocative occasions when they shared similar good experiences with their own children or grandchildren (see Boxes 3.3 and 3.4).

Box 3.3 Vivid and multisensory childhood experiences of natural environments: quotations from urban communities in central Scotland

Q. What did you do when you were small?

'Collected conkers, look for fishing in the river; there's hardly any fish left now.'

'I was always in Greenfield when I was a wee lassie, climbing the trees.'

'We used to cook just at this little dip, and we used to play in it [the Water of Leith, an urban stream]. We use to get the swimming trunks on and we used to swim . . . It was very wild.'

Teenagers and adults

Q. What do you like to do now?

'My brothers like to make dens with friends, up in the woods in Corstorphine Hill.'

Teenager

'There is plenty for the kids to do . . . throwing stones in the river, climbing trees and making pirates' boats out of trees that have slightly fallen down. [On] some of the dead trees, you can actually hit the stick and make what we call a music tree . . . get music out of it.'

'Another family and we got together and we decided we would take them in the dark walking, and it was December . . . to make it feel how it was like when you were in the dark, there were no lights along the [disused] railway.'

'I tell you one thing that stuck in my mind . . . I allowed them [son and friends] to take a drink out of the burn, and none of them had ever taken a drink out of the burn [stream]. I couldn't get them away from it: 'Can I get another drink? . . . Can I get another drink?''

Parents and carers

3.3 Playing outdoors with responsive materials: Qingdao, China.

Source: author

3.4 Playing outdoors with responsive materials: Edinburgh, Scotland.

Source: author

> **Box 3.4** Vivid and multisensory experiences of natural environments: quotations from communities in the English East Midlands
>
> *'Years ago I remember taking my grandchildren, a number of children, walking around where the herons were. It was amazing because the old railway line had grown over and we saw some absolutely beautiful butterflies down there. It was beautiful and as we went over one fence there was a jackdaw sitting there . . . and I said 'good morning' to it and [laughs] it said 'good morning' back as it had obviously been someone's pet. It was so funny.'*
>
> *'The most wonderful feelings in nature are at dawn and dusk . . . magical things happen then.'*
>
> *'I've never met a group of kids from, say, five to ten who don't thoroughly enjoy themselves getting wet and mucky outside, they just love it. It's getting them there in the first place and turning them free to get their hands dirty.'*
>
> Parents and teachers

What is even more striking is the evidence from quantitative analysis of questionnaire data coming from reasonably large samples in Scotland (n=336) and England (n=459), which shows that the frequency of childhood visits to green spaces or natural sites such as woodlands is a powerful and highly significant predictor of how often people will visit such places as adults (Ward Thompson et al., in press).

In the Scottish data, use of logistic regression models to predict how often respondents visited woodlands showed that frequency of childhood visits to woodlands was by far the best predictor, with a success rate of over 71% (Ward Thompson et al., 2005). The models were most effective in predicting those who were unlikely to visit woodlands more than once a month, if ever, where the success rate was over 83%. The frequency of childhood visits was the single highly significant factor out of all the demographic and background variables measured in the questionnaire.

In the East Midlands data from England, frequency of childhood visits to green spaces was again a significant variable (one of three significant demographic and background variables) in predicting how often people visit as adults (Bell et al., 2004). Logistic regression models using childhood visits and one other variable (categories of who, if anyone, accompanies the person on a visit to green spaces) had a success rate of over 75% in predicting those who visit green spaces more than once a week compared with those who only visit once a year.

There are also some very interesting and statistically significant correlations between childhood visits to natural places and people's attitudes to such places as adults. The Scottish experience shows that only those who visited woodlands at least once a week in childhood are likely to go walking alone in woodlands as adults. The East Midlands data shows that those who visited green spaces at least once a week as children are much more likely to be users of green space within walking distance as adults. Only frequent childhood users of green space are likely to feel energetic in green spaces as adults, or to think of such places as potentially 'magical' or with some transcendental quality.

Such analyses show that there is an important relationship between childhood access to nature and adult habits in visiting woodlands and other green space. In the light of other research that demonstrates the physical and mental health benefits of access to green and natural environments (e.g. Bird, 2004) these findings assume an even greater significance. Pretty et al. (2005) have summarised a growing body of evidence that engagement with green spaces and nature affects health, categorised according to three levels of engagement: viewing natural environments; being in the presence of nearby green space or nature; and active participation and involvement in nature, for example through walking. Our research suggests that health concerns about the lack of physical activity and levels of mental stress and depression prevalent in adults today need to take into account the role of childhood experience in relation to outdoor and natural environments. It may prove much more difficult for adults to change to healthier lifestyles and obtain the benefits of access to natural environments if they have not had frequent childhood experience of such places.

CATHARINE WARD THOMPSON

Teenagers and engagement with the landscape

We have underlined the importance of early childhood access to the landscape but what about adolescents and the teenage years? OPENspace has gathered evidence from both local and international studies to explore teenagers' experiences of outdoor places, and the factors that are important to them. Despite the different attitudes and needs that arise from different age groups, there is a surprising level of commonality across Britain and North America, perhaps reflecting a transatlantic culture shared especially by teenagers (Travlou, 2003). The importance of the environment in supporting a developing sense of self-identity and independence in teenagers is reflected in their desire for places that support their social lives. Adolescence is a stage when young people are very much focused on themselves and their peers, which means that other age groups are rarely commented on except when, for example, older teenagers or adults interfere with their sense of freedom. Teenagers want both places to be comfortable with friends and places to 'be oneself', or on one's own; in either case a sense of ownership (figurative rather than literal) and a lack of oversight or intrusion by others is important.

It is particularly noticeable that, whether they come from Edinburgh or New York City or Sacramento, California, teenagers describe the outdoor and public places they like in terms of their social characteristics (what friends they meet there, how they can 'hang out' without interference), while descriptions of places they dislike are very frequently described in terms of their physical attributes, for example 'dirty', 'uncared-for', 'smelly', 'full of litter'. This suggests that physically attractive places are important to teenagers but not sufficient to engage their interest unless they also afford relevant social opportunities and support. By contrast, where places are highly socially attractive, the physical environment may be of little moment.

The social focus of their lives means that many of the places that teenagers older than 14 years choose to visit are related to commercial contexts (cafés, fast-food outlets, shopping malls) where there may be crowds of other people but a concomitant sense of anonymity and freedom that may not be available at home. However, for 12–14-year-olds, often still interested in making dens, climbing trees and other playful activity, wild and natural open space offers important opportunities for unregulated and adventurous play. For older teenagers, nearby green space such as woodlands can also offer attractive environments for places to 'hang out', by contrast with public urban spaces where they are often unwelcome (Bell et al., 2003; Ward Thompson et al., 2006) (see Boxes 3.5 and 3.6).

Our work for Natural England on wild adventure space for young people showed that unstructured use of comparatively wild or unmanaged space such as derelict sites, urban fringe woodland and other marginal areas can be particularly important to teenagers for social as well as physical activity, although we know comparatively little about such use beyond the anecdotal (see Box 3.5). For young men, in particular, some natural environments can provide the opportunity for tests of physical skill and daring, ranging from off-road motorbiking to building structures in the woods; being able to access such places without jeopardy may be a key contribution to social development. Natural environments further afield may offer more exciting and challenging opportunities, including motor sports and water sports, but these are less likely to be accessed by teenagers on their own. As with other age groups, for woodlands and green places to be used by young people every day, they need to be close to home (Ward Thompson et al., 2006).

Older people and access to nature and the outdoors

It is perhaps self-evident that natural environments can offer playful and exciting opportunities for children and young people. There has been less emphasis to date on access to nature for elderly people but OPENspace's developing work with this age group points to their enthusiasm for getting outside.

Our surveys in central Scotland showed that people over 64 years were some of the most positive about feeling at home in woodlands, more so than any age groups under 45 years, and they differed significantly from any under-45s in their rejection of the idea that woodlands might be fearful places, concealing unwelcome or potentially threatening strangers. Over-64-year-olds were the most interested of any age group in visiting woodlands with parking facilities (presumably they feel less able or willing to walk to local woodlands) and liked having a variety of

3.5 Making dens in local woodlands, Scotland.

Photograph by Katherine Southwell, © OPENspace

3.6 What teenagers like – a place away from adults 'where you can do what you want', Scotland.

Source: author

Box 3.5 Young people's attitudes to outdoor places in different parts of England

Q. What would you choose to do outdoors?

'Making dens down at the old railway track – all day'
  11/12-year-old, Burton Green, Warwickshire

'Free to do your own thing . . . No restrictions when you are outside'
  13/15-year-old, South Ockendon, Essex

Q. What would your ideal place be?

'Warm, with shelter, light, a nice big field, away from houses, within 20 minutes' walking distance, with comfy seats, no adults. A place where you could do what you want.'

'It would have been nice to have a place where we could enjoy ourselves without having our parents around, a kind of natural place where we wouldn't need to take the bus and spend money on fares and thus rely on our parents'

'The council and local authorities should recognize that what all teenagers really want is a warm and quiet place to hang out, so if open green spaces could provide this warm place without adults, they would become so popular'

'Teenagers don't really want to be on the streets, they want to be somewhere with their friends where is no one to tell them to get off'
  Young people, 12–16 years old, Coulby, Cleveland and South Ockendon, Essex

trees to experience. Many were positive about family walks, often including grandchildren, but they also recognized challenges of accessibility (see Box 3.6), particularly for carers of children.

> **Box 3.6** Older people's attitudes to visiting woodlands: central Scotland and English East Midlands
>
> *'I've walked with the Ranger at Cammo quite recently and he explained about the trees and everything about there, which was very interesting. I've also taken up cycling after years of never having been on a bicycle, and I've been on [Edinburgh cycle paths] . . . it was very pleasant, very nice.'*
>
> *'I use [the woods] three days every week when I have got the grandchildren. The size of the gate . . . to get a pushchair through, it's extremely difficult, it usually takes a man's strength rather than a woman's. Once you do get in with a pushchair, the paths are about 100 yards, steps, then it drops 10 inches and moves along and drops 10 inches, and it's fine down the way but a mother with a young child will have great difficulty getting the pushchair back up again.'*
>
> Over-60s, Central Scotland
>
> *'I think we need the accessibility for people to see some of the things. The youth have no problem getting anywhere but . . . I can only comment on one particular walk that I do now which at one time was lethal, you did it at your own risk: going from Matlock to Matlock Bath through the wood, up St John's Road. I thank whoever did the steps and the availability, because that just makes another pathway to wonderful nature, and you can stand in that wood and even though the road is so close it is so silent and that to me is a joy.'*
>
> Older person, East Midlands

Other research has reinforced the evidence that older people are clearly very active in getting out and enjoying natural surroundings but, as age increases, the challenges and barriers become greater. Focus group discussions with older people in urban, suburban and rural areas of Britain (I'DGO, 2005) show that getting outdoors is associated with a better quality of life, which encompasses concepts of independence, an active social life, good health and good neighbourhood environments. Meeting other people is one of the most enjoyable things about getting out for older people, demonstrating perhaps a surprising commonality with teenagers in the importance of the social dimension of getting outdoors. Participants also frequently mentioned enjoying fresh air, walking, feeling healthy and enjoying the scenery. What they dislike are bad pavements and difficult road crossings, lack of benches and poor access to toilets, and they often mentioned fear of crime and of young people's behaviour (see Box 3.7).

> **Box 3.7** Older people's attitudes to getting outdoors: Edinburgh, Glasgow and rural Cornwall
>
> Q. What do you like about getting outdoors?
>
> *'Quality of life to me is being able to go out, walk about and see things.'*
>
> *'I think the pleasure is meeting with people when you are out.'*
>
> *'You'll go for a walk, and you feel better when you come back.'*
>
> *'There is something aesthetic about going out.'*
>
> Q. What do you dislike, what are the barriers to getting outdoors?
>
> *'The pavements are dreadful. Absolutely dreadful.'*
>
> *'If I see four or five boys coming toward me, I panic. You can pass them and they make rude remarks and things like that, and you're quite frightened.'*
>
> *'When they wear trainers, you can't hear their footsteps.'*
>
> *'You don't hear the bicycles coming.'*
>
> *'Lack of benches. [You need] benches here and there to have a rest and to sit down.'*
>
> People over 65

PLAYFUL NATURE

The results of a more extensive survey of older people's experience confirm that access to outdoor environments adjacent to where people live plays a significant role in their quality of life (Sugiyama and Ward Thompson, 2007 and Chapter 10). One aspect of a good quality of life for many older people is being able to maintain healthy levels of walking, especially within their local neighbourhood. Good pavements and an attractive route to neighbourhood open space appear to be an important inducement to walking. Good facilities in the local park or open space, such as a café and toilets, are also significantly associated with walking for transport by older people. However, when it comes to recreational walking, that is walking for pleasure or leisure, the significant factors turn out to be lack of nuisance (no dog fouling or problems with youths hanging around) and pleasantness of the open space in the neighbourhood. The 'pleasantness' factor includes aspects such as its welcoming and relaxing atmosphere and its suitability for chatting with people and children's play, as well as the quality of its trees and plants, reflecting social as well as aesthetic considerations (see Figure 3.7).

It is clear, therefore, that engagement with the natural environment for older people offers a range of positive opportunities, including healthy activity, maintaining social networks and the chance to be with children in a playful environment. However, the findings also suggest that young people 'hanging around' in open space are often a deterrent for older people. What are the issues associated with intergenerational use of outdoor environments?

3.7 Good paths, attractive vegetation and the chance to meet people – a good place for people of all ages

Photograph by Katherine Southwell © OPENspace

CATHARINE WARD THOMPSON

Intergenerational issues: teenagers and older people

While teenagers and older people share a liking for the social contacts and support that outdoor environments afford, it appears that teenagers pay little attention to older people unless the latter actively intervene in the teenagers' activities. By contrast, older people are frequently nervous about aspects of young people's behaviour, often simply because their own reduced vision, hearing and/or physical robustness make them more vulnerable to being caught off-balance by a fast-moving young person coming past unexpectedly. This creates a challenge for designers of public space; respect and empathy for another person's, perhaps very different, outlook and physical state is a necessary component of sharing public space but one to which adolescence and youth seem particularly poorly adapted. As recent legislation in Britain covering Anti-Social Behaviour Orders is used to exclude teenagers from public space – a space important to them for their social and physical development – there is a challenge to find ways that allow shared space to function well and enhance positive intergenerational contact (see Figure 3.8).

Of course, young people of different ages and socio-cultural groups have different attitudes and behaviours. In general, younger teenagers have a positive and playful engagement with natural environments but older teenagers and young adults appear to go through a phase of disengagement with nature, when outdoor environments appeal more for what they offer in terms of social contact. In the sample of urban communities in central Scotland, the largest group of daily visitors to local woodlands was teenagers under 18 years old, and the largest group of weekly visitors was people aged over 64, suggesting a common interest in woodlands despite many differences in lifestyle. The largest group not visiting local woodlands at all were the 18–24-year-olds. A study in Southeast Hampshire (Leisure Industries Research Centre, 2001) also showed that young people's participation rates in outdoor activities are actually higher than those of adults but that there is a significant fall off in the use of managed countryside in the transition between childhood and adulthood.

Models and methods for understanding people's engagement with the outdoor environment

Our work on people getting outdoors has built on understandings of people's dynamic and transactional engagement with place (Canter, 1985; Scott, 1999). We have also taken into account social cognitive theory (after Bandura, 1986), which suggests that people's activity patterns are influenced by individuals' recognition of opportunities for activities, their own

3.8 The challenge of sharing spaces between generations, Edinburgh.
Source: author

skills to conduct them, and expected benefits from them. A key concept in social cognitive theory is that of self-efficacy – a belief in one's ability to perform a particular activity in a particular setting. This has an inward focus on the individual. By contrast, 'environmental support' conceptualizes interaction at the level of a physical setting, such as a neighbourhood park or local woodland, and looks at how environmental factors can act as either barriers or facilitators to outdoor activity. We have found this concept of environmental support to be particularly useful in exploring how differences in the physical outdoor environment can make a difference to people's quality of life, either directly or by mediating the ability of people to undertake outdoor activities (Sugiyama and Ward Thompson, 2007b).

In order to explore environmental support, we have been using an approach based on Little's concept of Personal Projects (Little, 1983). An attraction of using this kind of constructivist method is that it treats all participants as co-researchers, starting from a premise that empowers and engages with each person and their daily lives. It allows us to explore each person's particular planned or desired activities and how the environment makes it easy or difficult to carry out each activity. It may help us identify the kind of environments, the qualities of places, that serve common functions well for all people, those places or qualities that support the projects of some people only, and those that serve the idiosyncratic projects of a single individual. This in turn may help us understand better the nature of affordances offered by environment – the potential for activity or social or emotional engagement that different outdoor or natural settings offer – and how such affordances are actualised or realised in practice by different people (Gibson, 1979/1986; Heft, 1997; Kyttä, 2006).

Places we like and places we don't

It has become clear from our many different projects that what is attractive about outdoor environments and what people like about being in such places is to do with things very different from what people dislike: the one is not the opposite of the other. For example, teenagers and older people have highlighted enjoying the social aspects of getting out but disliking physical aspects of the environments they might encounter. This is reinforced by work in personal construct psychology which shows positive and negative affect to be two different constructs; what makes people happy is not the opposite of what makes them sad. This means that we need to be wary of arriving at too simplistic conclusions about what people will do on the basis of attitudes expressed in focus groups and interviews. One example in our work illustrates this well: a woman over 65 years old, who lives alone on a peripheral housing estate and has to use public transport to get into town. Her immediate response, when asked what she disliked about going outdoors, was 'fear of crime', on which she expounded at some length, and one might assume from this that she would be very wary of going out at night into the city. Subsequent discussion, however, revealed that she frequently goes into the city, taking the bus on dark, winter evenings, because going to music and theatre performances is sufficiently important to her. The attraction of such projects clearly outweighs the detraction of contexts where this person might feel vulnerable to crime.

Thus we need to be careful in the interpretations of our findings and 'unpack' the meanings behind apparently straightforward statements. The ultimate test of whether we have got our research and our planning right is people's behaviour – what people actually do – in a given context in the real world. For planners and designers, therefore, the need is for good evidence about what kinds of environmental support are most effective for healthy and enjoyable or socially fulfilling outdoor activity.

Conclusion

Research demonstrates that there are many aspects of outdoor environments and green spaces that are attractive to people, regardless of age. Certain natural settings are particularly potent in eliciting pleasurable responses, offering people opportunities for engagement with their environment on a range of levels. Mature woodlands are one such setting and beaches or sea-shores are probably another, although we have less research evidence available on them (are sandy beaches the archetypal playful environment?). The constraints to visiting such places vary rather more for different social and cultural groups and we need to understand better the way in which environments offer opportunities or deter different individuals and groups.

The key significance of playful engagement with nature and open space for children is underlined by the way that it appears to resonate with people for the rest of their lives. Free and easy access to adventurous and enjoyable outdoor environments is important: it has the potential to confer a multitude of benefits on young people's physical development and wellbeing, emotional and mental health, and societal development (Ward Thompson et al., 2006). Benefits from play in natural settings appear to be long term, realised in the form of emotional stability in young adulthood and meaningful engagement with outdoor environments in later life (Travlou, 2006). Thus restrictions on children's experience of the natural environment is constraining on subsequent attitudes, behaviours and patterns of life, with consequences for society as a whole. Recent evidence also points to the serious consequences of restrictions on such experiential play for children's cognitive and conceptual development; in Britain, 11- and 12-year-old children have been demonstrated by Shayer (Crace, 2006) to be between two and three years behind their counterparts 15 years ago, with significant implications for the next generation. In the light of this, we need to ask serious questions about the way that access to the environment has become more restricted and controlled for children of all backgrounds in recent decades (Valentine and McKendrick, 1997) and consider what the consequences will be for healthy lifestyles.

We also need a more sophisticated response to young people's need for access to open space as they go through adolescence and approach adulthood. Teenager-friendly policies for the public domain cannot be developed while society is still employing crude stereotypes of children and young people as either victims or incipient miscreants (Worpole, 2003). Conceptualization of young people as a problem and a threat is widespread and has contributed to their marginalization and social exclusion. This is particularly relevant in the case of urban outdoor environments – the places to which most young people have easiest access in our urbanized society.

Ultimately, the goal must be access for all ages to environments that are rich in opportunities and support for health, development and wellbeing. Responding to the playful natures in all of us by creating good access to natural environments is one, vital way forward and the evidence is mounting that society cannot afford to ignore such demands.

References

Bandura A. (1986) *The Social Foundations of Thought and Action: A Social Cognitive Theory*. Englewood Cliffs, NJ: Prentice-Hall.

Barnard, J. (2006) 'The nesting instinct', in *The Guardian*, Saturday 15 April 2006.

Bell, S., Ward Thompson, C. and Travlou, P. (2003) 'Contested views of freedom and control: Children, teenagers and urban fringe woodlands in Central Scotland', in *Urban Forestry and Urban Greening*, vol. 2 (2), pp. 87–100.

Bell, S., Morris, N., Findlay, C., Travlou, P., Gooch, D., Gregory, G. and Ward Thompson, C. (2004) *Nature for People: The Importance of Green Spaces to East Midlands Communities*. English Nature Research Report no. 567. Peterborough: English Nature.

Bingley, A. and Milligan, C. (2004) *'Climbing Trees and Building Dens'. Mental health and well-being in young adults and the long-term effects of childhood play experience*. Research Report, Institute of Health Research, Lancaster University.

Bird, W. (2004) *Natural Fit: Can Green space and Biodiversity Increase Levels of Physical Activity?* RSPB.

Burgess, J. (1995) *Growing in Confidence: Understanding People's Perceptions of Urban Fringe Woodlands*. Northampton: Countryside Commission.

CABE Space (2004) *The Value of Public Space: How High Quality Parks and Public Spaces Create Economic, Social and Environmental Value*. London: CABE Space.

Canter, D. (1985) 'The Road to Jerusalem', in D. Canter (ed.) *Facet Theory: Approaches to Social Research*. New York: Springer-Verlag, pp. 1–13.

Cole-Hamilton, I., Harrop, A. and Street, C. (2001) *The Value of Children's Play and Play Provision: A Systematic Review of Literature*. N. P. Institute.

Countryside Agency (2004) *Diversity Review – Options for Implementation*. Research Note CRN 75, Cheltenham: Countryside Agency. Available at www.countryside.gov.uk (accessed 3 August 2006).

Crace, J. (2006) 'Children are less able than they used to be', *The Guardian*, Tuesday, 24 January 2006.

Department for Communities and Local Government (DCLG) (2003) *Sustainable Communities Plan*. Available at www.communities.gov.uk/index.asp?id=1139868 (accessed 26 August 2006).

Everard, B., Hudson, M. and Lodge, G. (2004) *Research into the Effect of Participation in Outdoor Activities on Engendering Lasting Active*

Lifestyles. English Outdoor Council. Available at www.reviewing.co.uk/research/active-lifestyle-obesity.htm.

Faber Taylor, A. and Kuo, F.E. (2006) 'Is contact with nature important for healthy child development? State of the evidence', in C. Spencer and M. Blades (eds) *Children and their Environments: Learning, Using and Designing Spaces*. Cambridge: Cambridge University Press, pp. 124–140.

Gibson, J.J. (1979/1986) *The Ecological Approach to Visual Perception*. Hillsdale, MI: Lawrence Erlbaum Associates.

Hansen, L.A. (1998) 'Where we play and who we are', in *Illinois Parks and Recreation*, vol. 29(2), pp. 22–25.

Hart, R. (1979) *Children's Experience of Place*. New York: Irvington Publishers.

Heft, H. (1997) 'Affordances and the body: an intentional analysis of Gibson's ecological approach to visual perception', in *Journal for the Theory of Social Behaviour*, vol. 19 (1), pp. 1–30.

I'DGO (2005) *Inclusive Design for Getting Outdoors*. I'DGO Project website available at www.idgo.ac.uk (accessed 28 August 2006).

Kjørholt, A.T. (2003) '"Creating a place to belong": Girls' and boys' hut-building as a site for understanding discourses on childhood and generational relations in a Norwegian community', in *Children's Geographies*, vol. 1 (1), pp. 261–279.

Kyttä, M. (2004) 'The extent of children's independent mobility and the number of actualized affordances as criteria for child-friendly environments', in *Journal of Environmental Psychology*, vol. 24, pp. 179–198.

Kyttä, M. (2006) 'Environmental child-friendliness in the light of the Bullerby Model', in C. Spencer and M. Blades (eds) *Children and their Environments: Learning, Using and Designing Spaces*. Cambridge: Cambridge University Press, 141–158.

Land Use Consultants (2004) *Making the Links: Greenspace and Quality of Life*. Scottish Natural Heritage Commissioned Report No. 060 (ROAME no. FO3AB01).

Leisure Industries Research Centre (2001) *Southeast Hampshire Young People's Countryside Recreation Demand Survey: Final Report*. A report by the Leisure Industries Research Centre, Sheffield Hallam University. Hampshire County Council.

Little, B. (1983) 'Personal projects: A rationale and method for investigation', in *Environment and Behavior*, vol. 15, pp. 273–309.

Lohr, V.I. Pearson-Morris, C.H., Tarnai, J. and Dillman, D. (2000) *A Multicultural Survey of the Influence of Childhood Environmental Experiences on Adult Sensitivities to Urban and Community Forests*. Department of Horticulture and Landscape Architecture, Washington State University. Available at www.wsu.edu/.

Moore, Robin C. (1986) *Childhood's Domain: Play and Place in Child Development*. Dover, NH: Croom Helm.

Pretty, J., Griffen, M., Peacock, J., Hine, R., Sellens, M. and South, N. (2005) *A Countryside for Health and Wellbeing: The Physical and Mental Health Benefits of Green Exercise*. Sheffield: CRN Countryside Recreation Network.

Richer, J. (2005) *Dirt is Good*. Available at www.smutsarbra.se/pressmaterial/Whitepaper.pdf.

Scott, M.J. (1999) '"Everything sounds like welcome home": community and place attachment', for British Psychological Society, *Social Psychology Conference*. University of Lancaster, September 1999.

Sobel, D. (1993) *Children's Special Places: Exploring the Role of Forts, Dens and Bush Houses in Middle Childhood*. Tucson, AZ: Zephyr Press.

Sugiyama, T. and Ward Thompson, C. (2005) 'Environmental support for outdoor activities and older people's quality of life', in *Journal of Housing for the Elderly*, vol. 19 (3/4), pp. 169–187.

Sugiyama, T. and Ward Thompson, C. (2007a) 'Older people's health, outdoor activity and supportiveness of neighbourhood environments', doi:10.1016/j.landurbplan.2007.04.002

Sugiyama, T. and Ward Thompson, C. (2007b) 'Outdoor environments, activity and the well-being of older people: Conceptualising environmental support', in *Environment and Planning A*.

Thompson, G. (2005) 'A Child's Place: why environment matters to children', in *ECOS*, vol. 26, pp. 9–12.

Travlou, P. (2003) 'Young People and Cities', paper presented in L. Maxwell (chair), Workshop on 'Teenagers' Experiences of Outdoor Places', *Environmental Design Research Association EDRA 34 Conference*, Minneapolis, 21–25 May 2003.

Travlou, P. (2006) *Wild Adventure Space for Young People: Literature Review*. Report to Natural England, Edinburgh: OPENspace. Available at www.openspace.eca.ac.uk (accessed 3 August 2006).

Valentine, G. and McKendrick, J. (1997) 'Children's outdoor play: exploring parental concerns about children's safety and the changing nature of childhood', in *Geoforum*, vol. 28 (2), pp. 219–235.

Ward Thompson, C., Travlou, P. and Roe, J. (2006) *Free-Range Teenagers: The Role of Wild Adventure Space in Young People's Lives*. Report to Natural England, Edinburgh: OPENspace. Available at www.openspace.eca.ac.uk (accessed November 2006).

Ward Thompson, C., Aspinall, P. and Montarzino, A. (in press) 'The childhood factor: adult visits to green places and the significance of childhood experience', in *Environment and Behavior*.

Ward Thompson, C., Aspinall, P., Bell, S. and Findlay, C. (2005) ''It gets you away from everyday life': local woodlands and community use – what makes a difference?', in *Landscape Research*, vol. 30 (1), pp. 109–146.

Ward Thompson, C., Aspinall, P., Bell, S., Findlay, C., Wherrett, J. and Travlou, P. (2004) *Open Space and Social Inclusion: Local Woodland Use in Central Scotland*. Edinburgh: Forestry Commission.

Worpole, K. (2003) *No Particular Place to Go? Children, Young People and Public Space*. Available at www.groundwork.org.uk (accessed 27 August 2006).

Part 2

The nature of exclusion: what is the experience of exclusion in different contexts?

Culture, heritage and access to open spaces

Judy Ling Wong

CHAPTER 4

Access to participation

Access by everyone to social goods is not a privilege but a right. The role of open spaces has won a significant place on the quality of life agenda. The agenda goes beyond provision. It is now recognized that access to the use and enjoyment of high quality open spaces, participation in their maintenance and improvement, and the creation of new spaces relevant to the needs of a range of social groups deliver a better quality of life.

Involving ethnic communities is one of the key challenges. Involvement allows open spaces to be created and managed in a way that is socially and culturally relevant. Within this essential context, there is a twin delivery of outcomes. Parallel to involving and benefiting members of any disadvantaged group comes the release of their vast missing contribution.

Multiculturalism and Britishness

Multiculturalism is not a result of social policy. Multiculturalism is a fact, accelerated within a world where communication and travel results in daily cultural interchange. Since people began to move across the earth, cultures mixed so that every culture is in itself multicultural. The description of a society as multicultural describes the ongoing meeting of cultures, with the 'dominant multicultural culture' having the most influence in shaping national life. It points out that the dominant culture is in continuity with the different cultural components embodied in its citizens. Society's multicultural character is expressed within each individual. Fish and chips is Jewish. We 'traditionally' celebrate Christmas with North American turkeys and potatoes. We count with Arabic numbers. We British would wither without millions of cups of Chinese tea. The romanticizing of the past has bred the myth of a pure culture. Within this vision we embrace hope for a harmonious nation and positive relationships with world cultures, something indispensable to the reality of global trade.

Every person who longs to enter Britain wants to come because they admire what it stands for and want to be part of what it is. I have never met anyone who wants to come here

JUDY LING WONG

with the aim of being separate. The separateness of different ethnic communities here is a result of racism, rejection and neglect. When hopeful migrants arriving here have their hopes dashed and are not welcomed as new British citizens, then they turn back towards what little they can personally carry from their cultures of origin. If one is not allowed to be British and accepted into mainstream society, then one has to be something, otherwise one becomes a nowhere person. Over decades of racism, obvious rejection and neglect, ethnic communities developed mini-cultures of their own.

However, look carefully and one will realise that none of them replicates the culture of their mother country. They are all British versions of aspects of the cultures of their origin. How can small numbers of people, usually from a very restricted social band, hope to recreate the swerving, heaving mass that national cultures are? Our ethnic communities are British already. Ask any person from an ethnic minority about their experience on visits to their home country, and they will tell you that they all feel like foreigners. Their countrymen there agree that they are – they are British.

To have a cohesive and integrated society, there is more for the mainstream population to awaken to and to do than there is for its ethnic minority citizens. The presence of immigrants, new citizens in cultural transition, and the richness of mutual impact is nothing new. It continues like it always has – all of us influencing each other, all of us taking what we think is good and rejecting what we do not like. The impact of different values sharpens the mind and accelerates positive change.

It is time for members of the dominant culture to stop blaming ethnic communities for their 'separateness'. Remove the rejection and racism, and there will be no more need for ethnic communities to turn towards themselves for security and social warmth. It is time for everyone to start taking responsibility to build social cohesion within the reality of a multicultural society.

The responsibility for social development

The present climate proposes that organizations and professionals have a responsibility to deliver social cohesion. The most direct expectation is the delivery of services and social goods that are relevant to a range of people defined as socially excluded. The most challenging aspect of working with any disadvantaged, socially excluded group is the necessity for attending to social development with respect to both the mainstream population and the social group concerned.

The straightforward reality of the existence of a facility or a service does not address the fact that various social groups are unable to use or benefit from them. Simple promotion or provision of information has been seen to fail to engender change. The complex picture of social relationships, or the lack of them, presents itself. Since the term social cohesion was coined by government, associated issues of equality of opportunity, cultural identities, the interpretation of history and heritage, social needs, racial discrimination and citizenship have been coming to the fore.

The evidence produced as part of the Diversity Review – 'Outdoors for All' – undertaken by the Countryside Agency for DEFRA (Department for Environment, Food and Rural Affairs) on Provider Awareness of the Needs of Under-represented Groups (Countryside Agency, 2005a, 2005b) concludes that currently the dissemination and communication of policies, initiatives and strategies is poor, and that there is insufficient and inadequate monitoring and evaluation. It recommends training and awareness programmes, guidance on the use of language, and the facilitation of effective, confident, sensitive communication between service providers and under-represented groups. A forum for discussion and events to publicize research, policies and best practice are also recommended.

Although this situation is increasingly recognized, professionals and agencies within both the statutory and voluntary sectors that have focused remits have logistic difficulties with respect to whose responsibility it is to work for social change and where the resources for long-term investment in community development will come from.

Furthermore, these issues cannot be approached in isolation. Issues of diversity and social inclusion are integral to all actions and can only be addressed as part and parcel of the facilities and services concerned. This means demanding that professionals' entire organizations that have hitherto only concentrated on discrete themed areas of work are suddenly challenged with respect to their working culture and asked to advance into being instigators of social change. The complex

CULTURE, HERITAGE AND ACCESS TO OPEN SPACES

skills and knowledge of community development need to be added to ring-fenced specialist skills that are the focus, the clear remit of organizations. These demands to link and integrate working themes, and to develop ways of doing so, chime in strongly with the present thrust for working in the context of sustainable development, anchored in the principles of putting structures into place for truly sustainable communities.

A partial solution is for organizations with focused remits to work in partnership with organizations that have been constituted with a more generalist agenda, for example those that are set up to work for the overall welfare of a local community. Sometimes this can work well, but at other times it can be cumbersome or impossible. This way of working can also inhibit client groups from benefiting fully from mainstream services, facilities and expertise. Such situations have indeed led to committed organizations such as BTCV (British Trust for Conservation Volunteers), an organization focused on environmental volunteering, considering changing their constitution in order to open up their organization's work programme to address the needs of disadvantaged groups alongside bringing forward their missing contribution (Davy, 2005).

Foward-looking professional organizations such as the Chartered Institution of Water and Environmental Management (CIWEM), through their 'Diversity, Society, Citizenship and the Environment' initiative, have made efforts to encourage professionals to consider and take steps to raise their own awareness and take individual action to take account of social and cultural needs as appropriate within their work.

Government departments, local authorities and statutory agencies are beginning to consider following the leadership of visionary organizations and strategically put into place diversity champions as one of the ways of giving a focus to addressing the complexities of true involvement of socially excluded groups (Department for Constitutional Affairs, 2004).

Case study: BTCV and BEN

Over many years, Black Environment Network (BEN) has been pro-active in putting ethnic participation on the agenda of the environmental sector. It has championed the reality of the need for the development of new ways of working within the environmental sector in order to establish full ethnic environmental participation. A significant part of this work involves enabling access by ethnic groups to a range of open and green spaces such as parks and gardens, neighbourhood green spaces, National Parks, urban woodlands, city fringe open green spaces, spaces associated with social housing and allotments (see Figure 4.1).

It has addressed issues including the use and enjoyment of facilities and services, involvement of ethnic groups in the care of these, and the creation of new spaces. Starting from where people are, the methodology developed was of necessity an integrated approach, bringing together social, cultural and environmental concerns. BEN proposes that there is no such thing as a pure environmental project, that a so-called pure environmental project is one that has rejected its social, cultural and economic context.

Environmental participation has been dominated by the white middle class in Britain. The analysis is that people whose lives are in order, can devote their energies to be a workforce for nature without asking for anything in return. Access to the enjoyment of open spaces such as National Parks or the open countryside is a normal part of their lives as they are well informed and have the resources to go wherever they wish. Whereas, the life situations of disadvantaged groups beset by problems necessitates the combination of being able to see environmental engagement deliver an impact on their lives at the same time that they make a contribution to the care and protection of the environment. Neither do they have the basis for motivation to care for spaces which they have never had the opportunity to enjoy. People simply do not know where to go, have no concept of what enjoying the countryside means and, in many deprived areas, there are few if any local green spaces or pleasant open spaces.

BTCV took on the BEN approach of starting where people are, finding a focus that delivers social, cultural, economic or environmental benefits that key directly into their lives while linking into environmental action that benefits nature. It has meant imaginatively linking people into opportunities that have a clear link to delivering their needs while opening up environmental activity. For example, as part of the Prudential Grass Roots project in Reading, BTCV identified a large, previously unused wooded area in the school grounds and created a

4.1 Young people enjoying Cashel Forest, Scotland.

fenced off area which has been transformed into an aquatic haven for frogs, newts and toads. BTCV staff worked with parents and children to create the pond and a raised vegetable bed for a gardening club. At the Greenpark Business complex, a nature trail has been designed as a fun and stimulating way to teach children about wildlife and the environment.

As part of their Environments for All programme, BTCV, in Glasgow, in partnership with multi-faith centres, has engaged a group of young Asians on a faith-based media project – the Positive Images project. After learning the skills of filmmaking, they are now using these skills to investigate the links between different faiths and the environment. The aim is to engage people from the different faith communities, for example Sikh, Hindu, Jewish, Muslim and Buddhist, in practical environmental activities reflecting their common beliefs. The project took them into a range of spaces new to them and stimulated interests in wildlife, laying down the basis for their future participation. Over eighty young Muslims studied the biodiversity within Mugdock Country Park using tree trails and treasure hunt activities. The Positive Images project illustrates the impact of innovative thinking and how new areas of interest and activity can be introduced by paying attention to the communities' needs and interests. The approach works, but the balance of investment in working to benefit people and working to benefit nature can shift to the point of an identity crisis for the organization – is it a community development organization delivering social benefits or an environmental organization working for nature?

The answer is that it is both. There is a dual challenge. Organizations such as BTCV hold enormous environmental knowledge and expertise to give impetus to the transformation and creation of open spaces linked to the theme of wildlife and nature. Focused on environmental volunteering, it needs people to act. While delivering environmental action, it takes people onto a range of green spaces, from local nature reserves and city fringe woodland to some of the most beautiful heritage landscapes of this country. Through its work with disadvantaged groups, it realises that, although the percentage of environmental action against the percentage of community development work done is much smaller than its work with members of the middle classes, it is nevertheless bringing forward a vast potential contribution that may never take place without the parallel community work. However, BTCV's decision on its direction of development is not its alone. The pressure of funders who focus their investments narrowly and their accountability to them is substantial.

Additionally, there is a cost attached to taking on such an approach. Training and organizational culture change, whether it is for independent professionals, the awareness raising of its board or managers, or the knowledge and skills of staff on the ground, costs in terms of time, resources and cash.

The challenge to a range of key players is change in terms of going beyond agreement on the need to address social change integral to themed work, but carried through to putting infrastructural support for change into place. The views of government departments and how they see their separate remits, the positioning of funders and the incorporation of training costs and work that crosses sectoral themes as legitimate items to be funded, investment of time by delivering organizations for organizational culture change, and the awakening of disadvantaged groups with regards to their rightful needs in relation to access to full participation as citizens – these are all part of the picture of development and change.

These questions seem to be gathering force around the concept of sustainable development. Sustainable communities, as the anchor to sustainable development, now mean the opening up of participation by everyone. The traditional narrow remits of focused themed sectors of concern – environment, social services, education and so on – are giving way, albeit slowly, to concepts of integrated delivery. The condescension of working in 'grey areas' will ultimately give way – the reality of sustainable development will see these so-called grey areas as the true golden areas of opportunity, the areas where the crucial but still legendary necessity of 'linking up' will happen.

BTCV's Environments for All project has blazed the trail for commitment to organizational culture change in the environmental sector. Black Environment Network continues to work in partnership with BTCV around innovative projects to strengthen and develop the way forward. Inspired by the impact of such work, key government departments such as the Home Office, with responsibility for social cohesion, have begun to recognize the potential of an integrated approach and have put funds towards the support of access to green spaces and environmental activities as a new area for the delivery of community development and social cohesion (Davy, 2005).

Multicultural interpretation and a sense of belonging

Black Environment Network coined the term 'multicultural interpretation' in the paper 'Multicultural Interpretation and Access to Heritage' (Wong, 1999). This paper was produced by BEN to coincide with the 'Whose Heritage' Conference in Manchester in 1999. It was commissioned by the Heritage Lottery Fund as a contribution to the debate on access to heritage.

A sense of belonging is central to citizenship. It is only as full citizens that members of ethnic communities can claim their right to access and full participation in the life of this country, and demand the visible recognition of its multicultural history.

The multicultural history of Britain has left a trail of features in our built and natural environment. Many existing features are evidence of multicultural history, of why many of our settled ethnic communities are here. Fountains and formal garden features inspired by Islamic cultures, pagodas and exquisite pavilions from China, monuments commemorating military events in Egypt, Africa or India, historic houses funded from the proceeds of slavery, or the large range of herbs in kitchen gardens and allotments are taken for granted: they are just there, as normal features that one encounters in the urban or rural environment. Much can be achieved through the multicultural interpretation of our urban and rural landscapes and their features. For example, some National Trust properties are the most multicultural properties in the world in terms of how the wealth created enabled them to be built, the origins of design and objects within the buildings, and the range of plants in the extensive open grounds. Explicit upfront interpretation that makes visible the multicultural history of Britain can go a long way to the re-positioning of mainstream identity in relation to the acceptance of the multicultural character of British culture

JUDY LING WONG

4.2 Al-Hilal mosque, community centre and garden. Manchester.

and heritage. The visible multicultural context of British culture and heritage enables a sense of belonging for ethnic communities. We are legitimately here because we are the consequence of a shared history.

> What we wish to do and what we can achieve depends on how we see ourselves against the enormous pressure of how others see us.
>
> Judy Ling Wong OBE, Black Environment Network (2002)

Coming closer to home, above all, it is the stuff of daily life in our immediate surroundings that makes all of us feel that we belong. It is about being able to walk safely in the local street, feeling like part of the local landscape. It is about growing up in the local park, kicking a ball around on a warm day, sitting around in the market square, picnicking in the National Park. The well-being of culturally defined disadvantaged communities remains oppressed by the condition and character of their surroundings against a context of racism and social exclusion.

> Perceptions of citizenship are being driven by a series of negatives (immigration, terrorism, etc.) that collectively highlight people's fears. Voluntary and community organizations can play a positive role in promoting community cohesion and generating social capital. Lack of understanding and/or tolerance of marginalised communities and their representative organizations represent a significant risk.
>
> The National Council for Voluntary Organizations (2006)

Social exclusion exists within a framework of exclusion. The 44 most deprived local authority areas in England proportionally contain four times more people from ethnic communities than other areas (Cabinet Office, 2000). People living in these areas named pollution, poor public transport and the appearance of the estate as major issues about where they live. The public realm, most notably our immediate environment, shouts messages at us. Run-down estates and neglected playing fields tell the local community that no one cares. Many young people have nowhere to congregate and have nothing to do. The elderly and lone women stay at home because they are frightened to be on the streets.

The lack of monitoring is indicative of the consistent neglect of the needs of ethnic minorities. There is a significant lack of information about minority ethnic groups in society, and about the impact of policies and programmes on them. But the available data demonstrates that, while there is much variation within and between different ethnic groups, overall, people from minority ethnic communities are more likely than others to live in deprived areas and in unpopular and overcrowded housing. They are more likely to be poor and to be unemployed, regardless of their age, sex, qualifications and place of residence. As a group they are as well qualified as white people,

CULTURE, HERITAGE AND ACCESS TO OPEN SPACES

4.3 Holi celebrations in the gardens of Shri Venkateswara Temple, West Midlands.

but some black and Asian groups do not do as well at school as others, and African-Caribbean pupils are disproportionately excluded from school. Pakistani, Bangladeshi and African-Caribbean people are more likely to report suffering ill-health than white people. Racial harassment and racist crime are widespread and under-reported, and not always treated as seriously as they should be. Minority ethnic communities experience a double disadvantage. They are disproportionately concentrated in deprived areas and experience all the problems that affect other people in these areas. But people from minority ethnic communities also suffer the consequences of racial discrimination; services that fail to reach them or meet their needs; and language and cultural barriers in gaining access to information and services (Cabinet Office, 2000).

People from ethnic minority backgrounds experience more health consequences from isolation and fear of crime in their local environment – instances of stress, depression, loss of appetite, increased alcohol consumption and lack of self esteem are consistently double in number compared to the population as a whole. Ethnic minority groups in general have lower levels of economic activity than white people. The employment rate for ethnic minorities in Great Britain in 2002 was 59%, as compared to an overall rate of 75%. The gap has been consistent for the last 20 years (Commission for Racial Equality, 2003). Overall, the ethnic communities have younger age structures than the white population and different ethnic groups experience inequality and disadvantage in education.

In many of the rural areas of England, Scotland and Wales, there are too few people from ethnic communities for them to form themselves into constituted community groups. They therefore have no public face and are extremely hard to reach and support. Often, they have no voice and no representation.

Open spaces are more accessible to ethnic minority children than any other location for leisure activity, but their satisfaction rates are lower, often related to fears over personal safety and racial abuse. Until recently, much research on themes significant to ethnic minorities excluded references to them, resulting in a lack of essential information to steer policy on many fronts. Unease over the issue of ethnicity often results in professionals adopting colour-blind attitudes that ignore ethnic and cultural differences altogether.

Open spaces present a particular opportunity. They are spaces where people repeatedly gather, linger, undertaking a range of activities.

Cultural festivals enable the expression of culture and a sharing of culture with the mainstream population. Northamptonshire County Council organizes a multicultural festival, the Roots Culturfest, in a country park every year. Hainault Forest welcomed local residents to celebrate Congo Independence Day.

New features can be commissioned to give recognition to local ethnic presence. Chumleigh Gardens in Burgess Park, London, has created a number of 'cultural gardens'. An area of the park has been set aside with raised beds for the use of community groups (see Figure 4.4).

JUDY LING WONG

4.4 Local community group growing food in raised beds at Chumleigh Gardens, Burgess Park, London.

Artworks with cultural associations can also be commissioned and designed into new spaces. The entrance to the planting area in Chumleigh Gardens is marked by a beautiful iron gate with motifs on the theme of nature and wildlife (Black Environment Network, 2005).

But purposefully working for the development of a sense of belonging does not have to revolve around concrete features that are culture specific. Through the management and development of activities we can enable a sense of belonging that is about marking a space with memory. The space remains itself but becomes transformed through the meaning of activities. If issues of conflict, anti-social behaviour, racism and safety are addressed within local spaces so that the everyday activities of playing in the street, cycling to the local shop, having an ice cream in a city square or going to the countryside can be taken for granted, then members of ethnic communities can continually mark and remark open spaces with the warm memories of communal life and feel that they belong.

Key considerations for policy and strategy development

Access and participation by ethnic communities in relation to urban or rural open spaces means enabling them to access the use and enjoyment of open spaces. As they move further along the road of involvement they will acquire an informed opinion about open spaces and begin to make a practical contribution to the care, improvement and creation of open spaces. All of this can precipitate particular social and cultural benefits for ethnic communities.

Opening up ethnic participation depends on the positioning of mainstream agencies responsible for the creation, care and management of open space, ranging from National Parks to managers of parks, civic spaces or neighbourhood playgrounds. The new cross-sector agendas draw in agencies that can use these spaces to fulfil other aims. For example, the positive impact of being outdoors on health and well-being, with opportunities to exercise or take up new interests and participate in communal life, is increasingly recognized, with the result that agencies with responsibilities for health and informal learning become involved with the potential offered within open spaces.

The challenge can be summarised as a challenge to the development of inclusive professional practice and organizational culture change that addresses social inclusion. Open space professionals and organizations that provide facilities and services need to ensure that they have the commitment and capacity to work effectively and in a relevant way with ethnic groups.

Organizational and professional development

There is a need to invest in awareness programmes and training to address organizational culture change, in particular building understanding and commitment at the top, so that those in power can ensure that policy is followed by action, supported by resources. Knowledge and skills for engaging effectively with ethnic communities can be built into the organization's professional working practice. Mentoring and developmental support for staff should feature as part of a framework to ensure their capacity to work confidently and effectively with ethnic groups. All of these need personnel time in terms of relationship building with ethnic communities. It is helpful if specific areas of work such as outreach form part of job descriptions. In the context of a history of neglect, nurturing a genuine working relationship based on trust can be challenging. A crucial part of this work should include the development of partnerships with organizations that represent ethnic communities. These representative organizations can act as facilitators in networking and building working relationships between open space organizations and ethnic community groups. Setting targets and monitoring in relation to access and participation by ethnic communities will enable an organization to track progress. Researching and promoting good practice is key to developing work with ethnic groups. As the work grows and matures, organizations will have case studies, innovative approaches and methodology of their own to share and contribute to building up a wider-ranging critical mass of good practice in relation to open space.

Partnership and dialogue

In relation to partnership, *equal partnership is an essential principle*. Power sharing in decision making, from consultation and participatory evaluation to having ethnic representation on key committees, is par for the course. The setting up of a forum for discussion should be considered. Ethnic communities and their representative organizations have knowledge and experience that is culture specific. Given the necessary support, ethnic community groups and individuals from ethnic communities have a contribution to make and can lead effectively. Often, better resourced mainstream organizations can play a role in building the capacity of ethnic groups to engage meaningfully with professionals and organizations through the provision of resources and opportunities to be exposed to mainstream processes, for example through inviting members of ethnic groups to observe board or management meetings, or buddying them with key members of staff. Providers of services and facilities should address barriers to participation jointly with ethnic groups. These include the cost of participation, the logistics and costs of transport and the lack of knowledge, information and experience in the use of a range of more challenging open spaces such as aspects of the countryside or National Parks. Issues can be highlighted to fuel debate. Dialogue and other forms of engagement with ethnic groups can be seen as opportunities to informally introduce knowledge of a range of job opportunities within the sector in order to build aspirations by members of ethnic communities to enter careers related to open space. Putting diversity champions into place within agencies related to open space is a good way to demonstrate commitment, in parallel with putting community-based open space champions into place, combined with capacity building and support for their work.

Funding policy

Many professionals and organizations find the burden of provision of diversity awareness raising and high quality training for embedding relevant policy development, managerial practice and community development skills across the organization difficult to resolve. This area of support has yet to arrive on the

JUDY LING WONG

agenda of relevant key funders in the statutory and private sectors. Funding policy has yet to recognize the significance of initiatives that combine environmental access and socio-cultural needs.

Design

Attention to the incorporation of multicultural interpretation in open spaces can maximize the role that the visible multicultural culture and heritage of Britain can play in the development of a sense of belonging. Being pro-active in the recognition of culture and heritage, when appropriate, in the design of open spaces can pay dividends. The Calthorpe Project in King's Cross used design to reflect the influence of different cultures in its building. The Project instructed their architects to reflect aspects of design from different countries and integrate them in spirit into a modern interpretation (Black Environment Network, 2005; Walter Segal Self Build Trust, 2006).

Consulting the community about their feelings about different areas can meaningfully identify local spatial problems. Often, design can be used as a solution for such problems. For example, Bankside Open Spaces Trust (BOST, 2006) used design to increase feelings of safety. A pleasant space with seating was not used because it had only one entrance. People felt that they could be trapped. The redesign of paths and the creation of a new exit encouraged people to use the space.

Multicultural interpretation

Recognizing the potential cultural symbolism of the elements of an open space is fundamental to multicultural interpretation. Recognizing and expressing culture and heritage does not have to mean the embodiment of obviously culture-specific elements in the form of concrete artefacts or design elements. For those that can relate memory and history to elements in space, a cultural vision is embodied. For those who cannot, it is just a space. Revealing the socio-cultural-historical meaning of an open space through multicultural interpretation can involve different elements and approaches. The presence of plants from all over the world embodies the multicultural history of this country. The movement of plants parallels the movement of people. Plant trails or simply labelling the origin of bedding plants in civic squares or traffic roundabouts can surprise most people. When Black Environment Network conceptualized the first 'cultural garden' – giving the children of the inner-city Walnut Tree Walk School in Lambeth the opportunity to choose plants from different parts of the world to represent the ancestral countries of the school children – we in fact ended up selecting plants that are typical of an English cottage garden! A horticulturalist colleague pointed out with humour that 'An English cottage garden is nothing but a collection of glorified foreign weeds'. One of the most inspiring outcomes was that once upon a time walking to school every day meant walking past what they thought were purely English front gardens. After putting the cultural garden together, they were excited to find that they had always been surrounded by plants from their countries of origin (Wong, 1997).

Site managers can seek out opportunities for introducing or shaping elements of an open space to recognize the local presence of ethnic communities. Lister Park in Bradford identified an area to create a formal Mughal Garden (see Figure 4.5), investing substantially in a feature that recognizes local ethnic identities (Black Environment Network, 2005).

The community has responded with a strong sense of belonging. It is their park. They enjoy the facilities and have programmes of activities such as healthy walking, with over 150 Asian women walking in the park in the early morning every day. Open space managers can also create opportunities within their programmes of activities to enable ethnic communities to express their presence through using aspects of an open space, for example by running culture-specific activities or events such as festivals. Enabling ethnic communities to undertake activities that mark an environment with memory can be a very attractive option for involvement. As part of its programme of creation of native woodland, Bestwood Country Park in Nottingham (see Figure 4.6) welcomed the Sikh community to use the woodland to celebrate the three hundredth anniversary of the birth of Guru Nanak (Black Environment Network, 2005). They also planted 300 new trees to create a new area of native woodland. The area is named Khalsa Wood.

4.5 The creation of the Mughal Gardens in historic Lister Park, Bradford, recognizes the local presence of ethnic communities.

4.6 Sikh community celebrating the three hundredth anniversary of the birth of Guru Nanak through planting 300 trees, creating Khalsa Wood as part of Bestwood Country Park, Nottingham.

JUDY LING WONG

Open spaces as settings for activities

Open spaces are wonderful settings for activities. They are significantly defined by the activities that they do or do not run. Pro-active initiatives, in particular outreach and consultation with ethnic communities in relation to what they would like, can be undertaken and opportunities created to enable socially excluded ethnic communities to make positive connections with open spaces on their terms. The development of relevant programmes of activities can enable the recognition and expression of the culture and heritage of ethnic communities. For example, many members of ethnic communities bring indigenous craft skills that can be shared.

The provision of group activities can be important. Here, attention needs to be paid to socio-cultural needs in the design of activities and services. For example, girls- or women-only activities, or a range of suitable foods in relation to religious requirements. When working with the first immigrant generation, translation needs in terms of information provision, or newsletters, or interpretation in relation to activities and services are significant. Such services do not always mean incurring costs. Some of these services can be negotiated through the goodwill of local ethnic community groups, through developing good working relationships with them. Groups can be encouraged to use open space for a range of activities. For example, open space can be used as an outdoor classroom for learning English or other skills.

The management of open spaces to ensure equality of opportunity for the enjoyment and use of these spaces, its programmes of activities and services, by a range of ethnic communities, is dependent on the successful provision of a sense of welcome as much as addressing issues of conflict between members of the dominant mainstream population or between different ethnic groups. In scenarios of conflict, ensuring a feeling of safety and the control of anti-social behaviour can sometimes be extremely demanding.

The challenge of opening up all that open spaces have to offer to ethnic communities will take effort and commitment. Success can sometimes be almost instant and dramatic. At other times, it can be a frustrating journey and a longer-term process. However, the overall picture is optimistic. The momentum for enabling full ethnic participation in open space in Britain is truly underway. It is cause for celebration. The fact that the Office of the Deputy Prime Minister funded Wong and Auckland of Black Environment Network to write *Ethnic Communities and Green Spaces – guidance for green space managers* (Black Environment Network, 2005), and sent a copy to the Chief Executives of over 800 local authorities across England is symptomatic of our times.

References

Black Environment Network (2005) *Ethnic Communities and Green Spaces – guidance for green space managers.* Black Environment Network publication. Available at www.ben-network.org.uk/resources/publs.aspx.

BOST (Bankside Open Spaces Trust) (2006). Available at www.bost.org.uk/.

Cabinet Office (2000) *Minority Ethnic Issues in Social Exclusion and Neighbourhood Renewal.* Available at www.cre.gov.uk/duty/reia/statistics_housing.html.

Commission for Racial Equality (2003) *Statistics: Housing.* Commission for Racial Equality. Available at www.cre.gov.uk/research/statistics_housing.html.

Commission for Racial Equality (2003) *Statistics: Labour Market.* Commission for Racial Equality. Available at www.cre.gov.uk/research/statistics_labour.html.

Countryside Agency (2005a) *'What About Us?' Diversity Review evidence – part 1. Challenging perceptions: under-represented groups' visitor needs*, by Ethnos Research and Consultancy, Countryside Agency, January 2005. Available at www.diversity-outdoors.co.uk/.

Countryside Agency (2005b), *'What About Us?' Diversity Review evidence – part 2. Challenging perceptions: provider awareness of under-represented groups*, by University of Surrey, Countryside Agency July 2005. Available at www.diversity-outdoors.co.uk/.

Davy, Clifford (2005), Director of Diversity, BCTV (personal communication).

Department for Constitutional Affairs (2004) *Minority Report – A Review of the Department for Constitutional Affairs' Diversity Strategy.* Available at www.dca.gov.uk/dept/minrept.htm.

National Council for Voluntary Organisations (NCVO) (2006) *Voluntary Sector Strategic Analysis 2005/06 – An overview of the operating environment and strategic drivers for UK voluntary organisations.* Available at www.ncvo.com/3sf/trends/?id=2276.

Social Exclusion Unit (2000) 'Minority Ethnic Issues in Social Exclusion and Neighbourhood Renewal'. Available at www.socialexclusion.gov.uk/downloaddoc.asp?id=114.

Walter Segal Self Build Trust (2006) *The Calthorpe Centre, London. A purpose built community building at Kings Cross*. Available at www.segalselfbuild.co.uk/projects/calthorpe.html.

Wong, J.L. (1997) *The World in your Garden*. Black Environment Network publications. Available at www.ben-network.org.uk/resources/publs.aspx.

Wong, J.L. (1999) *Multicultural Interpretation and Access to Heritage*. Black Environment Network publications. Available at www.ben-network.org.uk/resources/publs.aspx.

Wong, J.L. (2002) *Who We Are: A Re-assessment of Cultural Identity and Social Cohesion*. Black Environment Network Publications. Available at www.ben-network.org.uk/resources/publs.aspx.

Landscape perception as a reflection of quality of life and social exclusion in rural areas

What does it mean in an expanded Europe?

Simon Bell and Alicia Montarzino

Introduction

This chapter examines the situation of people living in rural areas and how quality of life is affected by environmental, social and economic factors. While most of the population of Europe live in cities, there remains a significant rural population who have problems similar in nature but different in the way they manifest themselves, which also need attention by researchers and policy makers. Central to this exploration is the perception of the landscape within which rural people live and the contribution this makes to quality of life.

In Western Europe, where industrialization and urbanization, in the main, was completed by the middle of last century, rural deprivation is often considered to be a small-scale and localized phenomenon (Spencer, 1997). Large numbers of affluent urban commuters often live in country villages and many people also retire to live on pensions obtained through well-paid work in cities. However, across the newly expanded Europe, in those countries of Central and Eastern Europe which joined the European Union in 2004, there are larger proportions of rural dwellers who often live in poverty, and the demographic changes taking place in those regions point to serious issues that need to be addressed (see below). It is interesting therefore to compare some of the common factors between the rural areas of richer and poorer EU countries, representing two opposing points on the spectrum, and to consider what quality of life means in these circumstances and how it differs from that of residents of urban areas.

This chapter starts by setting out the general socio-economic context of social exclusion in rural areas and then moves on to focus on two case studies where these issues have been explored in relation to the contribution of the the physical environment to the quality of life. The remote Highlands of Scotland and rural areas of Latvia, representing an affluent Western European and an impoverished Eastern European country respectively, are compared in terms of the interaction of the socio-economic factors and the landscape.

The socio-economic context for social exclusion

Even in the United Kingdom, currently the fourth most affluent country in the world, there remain problems for social inclusion in the rural areas of the constituent countries. For example, in England, the Commission for Rural Communities has identified three critical factors for rural people in general (Scharf and Bartlett, 2006):

- financial poverty, with disproportionate numbers of people who are unwaged, on low wages or small pensions;
- access poverty, meaning poor access to transport or social and other services;
- network poverty, where the venues and means of enabling informal contact and help from families and neighbours to take place are frequently lacking.

Among different social groups within the general rural population, two have been singled out for special mention: older people and young people. For older people, several specific problems have been identified (Commission for Rural Communities, 2006: 2):

- a lack of access to material resources;
- inadequate or poor quality social relations;
- lack of access to services and amenities; and
- disadvantage linked to rural community change.

In terms of young people, a recent report (Madgely and Bradshaw, 2006) identified a number of actions needed to improve the opportunities provided to young people in rural areas, including:

- increased support for residential places at further education colleges;
- increased provision of adequate transport options to enable young people to access post-16 opportunities;
- provision of information, advice and guidance for young people who are in employment (especially those in low-skilled, low-paid employment); and
- support for skills training and career opportunities to help those young people who want to stay in, or return to, rural areas, to help them develop careers in areas that have the potential for growth in rural communities.

In Scotland a significant part of the land area has been identified as being rural, defined by the Scottish Executive as areas where settlements have a population of less than 3,000. By analysing driving times to larger settlements, rural Scotland has been divided into two categories: 'accessible rural' – those areas with less than a 30 minute driving time to the nearest settlement with a population of 10,000 or more – and 'remote rural' – those with a greater than 30 minute driving time to the nearest settlement with a population of 10,000 or more (Scottish Executive, 2005). Analysis of people living in these areas has revealed major differences in quality of life, health, access to services and transport costs when comparing these two categories (Rural Poverty and Exclusion Working Group, 2001). In particular, the major concerns were identified as:

- Access: some services are unsuitable for local delivery, either because of diseconomies of scale or the nature of the service. Lack of access to certain services, such as transport and childcare, can also restrict access to employment opportunities.
- Higher visibility, which in small communities can aggravate poverty and social exclusion for those experiencing problems that carry a social stigma by inhibiting them from addressing their situation.
- A culture of self-reliance, especially among older people, which might exacerbate poverty and social exclusion if people are too proud to seek help.

In the United Kingdom a number of government policies and programmes have been devised to tackle these problems but this is not the case everywhere in Europe. In the countries of the former Soviet Union, such as the Baltic states, and the former eastern bloc which joined the European Union in 2004 or hope to join shortly – the so-called Central and Eastern European or CEE countries – immense social and economic upheavals have been taking place.

Since the collapse of the communist system some 16 years ago and within a context of a general decline in population, rural demography has been changing to the point where depopulation, accompanied by land abandonment, has become widespread (Westhoek et al., 2006). By way of illustration, the population of Bulgaria is declining by 5.9% annually, that of Latvia by 5.4% and that of Poland by 0.06%.

In Western European countries such as Portugal, northern parts of Sweden, France and Ireland, rural populations are also decreasing (Eurostat, 2006). In the United Kingdom, however, the rural population is generally increasing through the process of counter-urbanization, although not everywhere, and pockets of depopulation remain (Spencer, 1997; Stockdale, in press) and it is this influx of better-off people who mask the social and economic problems. However, in terms of the population trends, it is the CEE countries which are experiencing the largest demographic changes.

Young people leave the countryside to find work in towns and cities or to travel abroad to work, while the older people remain behind. Coupled with declining fertility rates, which are leading to a generally ageing population in CEE countries, this leads to a disproportionately ageing rural population. This disadvantage is added to the fact that rural levels of income tend to be lower, access to services (transport, shopping, medical and social care) is usually more limited and the quality of housing is often poorer than in urban or suburban areas. Conversely, the environment of rural areas may be better, being less polluted, with cleaner water, more nature (wildlife, natural habitats), less traffic, lower crime rates and a strong sense of community. People may grow a proportion of their own food and be capable of self-reliance and self-sufficiency within a network of community support well into old age. It is this higher quality of the physical environment and its role in offsetting some of the negative socio-economic aspects that is discussed in this chapter.

Researching engagement with the rural landscape

These issues are explored further by examining two research projects, one in Scotland and one in Latvia. The aim of both studies was to examine people's relationship with the place where they live and to attempt to illustrate how relationships with the landscape are fundamental to quality of life and reflect issues of social exclusion.

Scotland, as described earlier, has many remote areas and, while an affluent country, it nevertheless experiences problems associated with poor quality of life and social exclusion. Latvia, one of the former Soviet countries, one of the poorest and a new member state of the European Union, provides an example of the trends of rural depopulation and experience of rural poverty.

The two studies vary in scale, the Scottish example being of a single community, Strathdon, in Aberdeenshire, while the Latvian study covers six rural communities and has a much larger sample size. However, what makes them comparable is that in both cases the perception of the rural landscape was studied through a mix of qualitative and quantitative research methods. The qualitative approach used either individual interviews, focus groups or group workshops and the quantitative approach used a questionnaire survey that built on the results of the interviews and focus groups. In both cases the overall methodology was rooted very firmly in approaches developed from personal construct theory (Kelly, 1955) and Canter's Theory of Place (Canter, 1977). According to the latter theory, perceptions and values of landscape are considered to be constructed differently in a very personal way, depending on the interaction of three main factors: the physical world, the activities undertaken and the individual's beliefs. People's transactional relationship with place means that whether or not they value the landscape around them will depend on how it affects the way they live and their needs and desires in daily life. When exploring the contribution of the local landscape to people's lives, it is necessary to consider all three elements of place as identified by Canter, and the interaction between them.

Scotland: the Strathdon study

Scotland's rural population is 29% of the total of 5.1 million. The character of the highland landscape, being currently sparsely populated, has a history going back to the period known as the 'Highland clearances' when large numbers of people living in this part of Scotland were forced off the land by landowners

SIMON BELL AND ALICIA MONTARZINO

who wanted to use the land for sheep farming, which was considered to be more profitable than the low rents obtained from these small farmers. In a period stretching from the late eighteenth to the mid-nineteenth centuries, the people were gradually removed, firstly to coastal areas and later, especially after the potato famine of 1846, which affected Scotland as well as Ireland, many emigrated to Canada or other British colonies, or moved to the cities to work in industry (Prebble, 1963). The land remains to this day largely in a pattern of large, privately owned estates. These usually comprise a large mansion house, often with gardens and an ornamental park, tenanted farms on the better land, where cattle and sheep are raised, areas of managed forest, and vast areas of open hill and mountain used for sheep grazing, deer stalking or grouse shooting. Most people live either in individual houses on the estates or in small villages. Some people work on the estates and others service the local economy, for example as shop keepers, sub-postmasters or teachers, or else they run small businesses. In many areas much of the land is also owned by the state, which bought it from estate owners in order to afforest it during the twentieth century. This also provides some rural employment.

The study area of Strathdon is a large valley lying in the north east of Scotland, part of the Cairngorms mountain range, whose population is classified as 'remote rural' according to the Scottish Executive definition. It is typical of the highland landscape and economy described above. The area features a number of traditional Scottish estates, each of which contains a mixture of upland farming, forestry, deer stalking, grouse shooting and salmon fishing. It contains a number of small, scattered communities with a total population of some 400

5.1 A view of Strathdon, showing the farmland in the valley bottom, forest on the slopes and open hills above.

LANDSCAPE AND SOCIAL EXCLUSION IN RURAL AREAS

adults. The landscape contains some forest, part owned by the estates and part by the Forestry Commission, and also the potential for much more afforestation. Some members of the community are tenant farmers or estate workers. Tourism is presently a small element in the valley economy compared with many other parts of Scotland. The community is tightly knit and is involved in a number of community projects and initiatives.

There are also some strong historical traditions which create a sense of identity. One of these is an event called the 'Lonach'. This is a gathering of people in traditional Highland dress in August each year, based on the anniversary of a twenty-first birthday party for the son of a local landowner in 1822 (Casely, [2000]).

The research aims and objectives

The research into perceptions of the landscape and community (Ward Thompson and Scott Myers, 2003) was undertaken using individual, semi-structured interviews with members of the community who occupied different roles – landowner, forest manager, local farmer, entrepreneur, retired person and so on – and by subsequently carrying out a questionnaire survey of 47 people. The interviews were structured around seeking answers to four basic questions: what people like about Strathdon, what they do not like, what they would like to see changed and what they would like to see stay the same. In addition, a series of participatory workshops was held where members of the local community could discuss the landscape and how they identified with it and with special places.

Results

The results of the individual interviews were grouped into those issues relating to the physical environment and economic considerations which enable or prevent people from continuing to live there. The physical landscape around Strathdon emerged clearly as a defining feature of the village. People liked the remoteness of Strathdon but also being within reach of larger towns and amenities. They felt that road access could be improved to encourage tourists to visit, which would then generate income. However, this needed to be balanced with

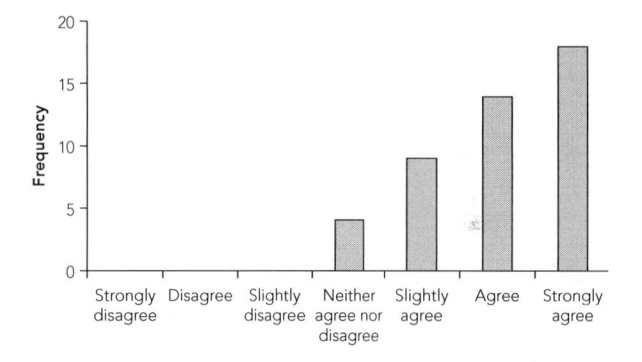

5.2 Graph showing the degree of agreement/disagreement with the statement 'The landscape of Strathdon is a reason for me being here'.

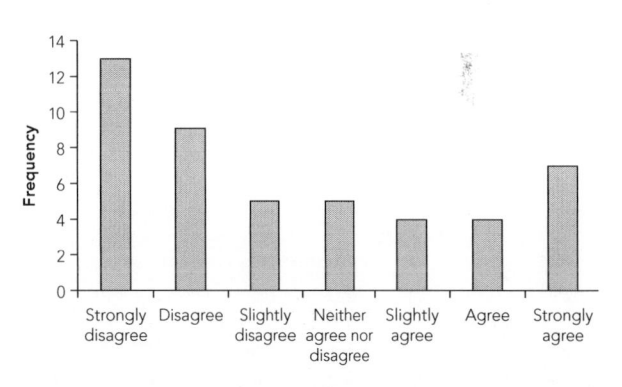

5.3 Graph showing the degree of agreement/disagreement with the statement 'I live in Strathdon to earn a living'.

maintaining the spirit of the community and was one of the central debates linking many aspects of the place evaluation: how to increase revenue while not spoiling the beauty and isolation of Strathdon or its community.

In general, the interviewees liked the physical setting of the village. They understood the economic demands and benefits of farming the land and the changes that happen when trees are grown as a crop. While the remoteness presented problems for getting to some amenities, this was generally seen as a trade-off worth making to access the perceived benefits of living in a remote place.

The social environment of Strathdon was represented in interviews by expressions of the strong community, the friendly people, their helpfulness and acceptance of different behaviours. The people and the community were considered the strength of the place and the reason that many liked to live there. One issue that may in future force integration between long-standing residents and 'incomers' is the economic state of the village. The population had decreased and the composition of the village had altered as some new people came in. Tourism was seen as a possible avenue for income, but this seemed to be a struggle, both in terms of getting support to develop tourist facilities and in the inevitable conflict between bringing visitors to Strathdon and maintaining the remote character of the place. Without new sources of income however, it seemed possible that Strathdon will fail to support itself, with the possibility that all the village services will disappear and with them the 'heart' of the village. For many, the accepted practices and traditions of the village were part of their personal identity and also what they believed Strathdon to be. This definition of self is well documented within psychology (Breakwell, 1993) and its relationship with the physical environment is also well established (Korpela, 1992; Twigger-Ross and Uzzell, 1996).

The overall implications from these findings are that the people of Strathdon are very attached to their village and many regard it as not only their current home, but also where they always wish to be. This is evidence of a very strong place attachment, and one that is likely to be influenced by the historical factors such as the tradition of the Lonach described earlier.

The questionnaire survey expanded on the central issues of the study, based on the results of these initial interviews. It revealed that nearly 50% of the respondents had lived in Strathdon for ten years or less, while 15% had lived there for 50 years or more; over 27% could trace back their family ancestry in the Strathdon area for more than 100 years. Farming and tourism accounted for less than 13% of the sample's income source, and 26% of the sample relied on more than one income source.

A number of attitudinal questions were developed in the survey, designed to cover aspects of the physical environment, people's activities and their perceptions or conceptualisations of their environment, drawing on what people had identified as important in the earlier interviews. Analysis of the survey data explored the interactions between different factors in people's responses. The landscape and its qualities of quietness, for example, were revealed as very important and contributing significantly to quality of life. Social issues centred around the importance of the community and its role in linking people together and helping them to support one another. The divide between incomers and locals was reflected in a mixed response to the importance of attachment to Strathdon through family ancestry. The economic issues were also important but it emerged that people live in Strathdon primarily because they want to, not because they need to – only 34% of the sample identified earning a living in Strathdon as a reason for living there.

The participatory workshops enabled some of the aspects of place attachment, identity with the area and its contribution to quality of life to be explored in more depth. The living environment emerged as the key value placed by people on Strathdon, especially the sense of community which, given the remoteness of the area, also helped to contribute to the sense of place. This is reinforced by the sense of neighbourliness and the fact that most people work locally or close by and there are no commuters. The fact that the land provides a livelihood for at least some of the community also reinforces the attachment. The landscape is an asset that could be exploited more through the development of tourism, for example. The participants also defined the sense of place as the river and the way it connects each part of the area and its symbolism as a source of purity. The wildlife, as an element of nature, was also valued and was associated with the expanse of moorland that forms the skyline around the strath. Once again, the social and physical environment emerged as important, linked aspects which together define the area and reinforce the attachment of people to it. People also recognized that what was good for one person's life was not necessarily the same for another person. For example, planting more forest or felling a mature area for timber could provide income for the landowner and a job for a local person but it might spoil the view from the house of another person who ran a bed-and-breakfast establishment. People recognized these trade-offs as part of the life of a community like Strathdon.

5.4 The moorland landscape is one of the defining features of Strathdon for many of the residents.

The Latvia study

Latvia has a rural population that constitutes 31% of the total population of 2.3 million (Earthtrends, 2006). To find the elements that have defined the character and economics of the rural landscape of Latvia it is necessary to go back to the Soviet era and to look at what happened after independence was regained. During Soviet times, all land was nationalized and farms were managed as collectives (*kolkhoz*), with large-scale mono-cultural production (Melluma, 1994). After regaining independence, the land was handed back to the previous owners or their descendants, many of whom lived in other countries following earlier exile, had moved to towns and cities or were not interested in farming, all of which led to the abandonment of many properties. Also, as people became free to leave the collective farms to which they had previously been tied, the population and economic structure of the countryside became fragmented.

The type of farm settlement and housing structure, as well as migration patterns and employment, have had an impact on rural living conditions, social structure and quality of life (Deller *et al.*, 2001; Kinsella *et al.*, 2000; van der Ploeg *et al.*, 2000). Traditionally, the prevalent type of farm settlement in Latvia was one of dispersed farmsteads with no concentrated village centre.

Before the first Latvian independence period of 1918–1940, the land was divided into a structure of large estates, much like those of Scotland, owned by Baltic German landowners who maintained manor houses in the 'villages', where there was also a church. These estates were broken up in the first

SIMON BELL AND ALICIA MONTARZINO

independence period and the countryside became a landscape of small- to medium-sized owner-occupied farms. Following the incorporation of Latvia into the Soviet Union and the collectivization of the farmland, populations were concentrated into blocks of flats in what became village centres (Lūse and Jakobsone, 1990; Grave and Lūse, 1990).

The collectivized farmland was reorganized into what were considered to be efficient units for mechanised agriculture, drained where necessary and the rest abandoned to revert to forest. As the majority of the people were moved into the new flats built for the collective farms, many of the older traditional houses were left empty and fell into disrepair. This has resulted in a population still now living in these flats, which were often of a poor construction quality, or living in the previously abandoned houses they have regained following land restitution but which are also in a poor state of repair and lacking modern services.

The research aims and objectives

Against this context, research was undertaken in six rural municipalities – each of a similar scale to Strathdon – and three urban areas, in different parts of Latvia, in order to uncover the values and perceptions people have towards the rural landscape and the quality of life available there, including the physical and social changes taking place. No such research has been undertaken before and there has been, to date, no basis for developing rural policies that take account of people's views. Focus groups were carried out in three rural locations and also in Riga and, on the basis of these, questionnaires were developed in a similar way to that used in the previous study. Some 30 persons from all age groups completed the questionnaires in each village or urban centre, enabling the difference between perceptions of young and old people, for example, to be explored.

5.5 A Soviet-era block of flats in a village in Latvia, typical of the housing built during the collectivisation period.

5.6 A rural scene in Latvia which is perceived as typical and which is associated with a sense of being Latvian, with birthplace and with home.

Results

In the focus groups the importance of traditional elements or aspects of the Latvian countryside became apparent. As part of the inventory of the Latvian landscape, interviewees in all groups consistently mentioned (in descending order of frequency) hay cocks, storks, detached farmsteads, thatched buildings, country bathhouses, old oak trees, avenues or rows of oak and lime trees, lakes, cultivated fields, country estates without hedges, winding highways, hillocks and flower gardens. Many interviewees also mentioned manor houses surrounded by old parks with ponds and nearby villages. The idea of the Latvian landscape was also linked with the places where their forebears lived, with childhood reminiscences and with feelings of home as well as with patriotism.

When comparing the landscapes of the First Independent Republic of 1918–1939 with those of both the Soviet period 1945–1991 and the restored Independent Republic from 1991 onwards, most people considered that the landscape has changed for the worse. They attributed this to the land abandonment, the increased forest cutting and the lack of maintenance compared with previous eras.

The interviewees tended to be rather pessimistic about aspects of the future, such as employment and the survival of the traditional ways of life, and they were especially worried about the trend for younger people to move away from the countryside to the towns and abroad. This did not, per se, imply a rejection of the rural way of living but was, rather, related to factors such as job opportunities. Many people wanted to continue to live in the countryside as long as they could work in a nearby town.

The questionnaire survey enabled the themes uncovered by the focus groups to be asked of a wider sample of people. Part of this questionnaire asked interviewees to provide up to ten words that came to mind when they thought of the Latvian countryside. This revealed a very strong dichotomy in the perceptions. On the one hand, there were very positive views of the countryside in general, the physical environment and its traditional character associated with community, tradition and neighbourliness as well as the qualities of the landscape such as nature, quietness, clean air and a pure environment. On the other hand, they also associated the same countryside with negative social and economic aspects, such as unemployment, poverty, hard work and alcoholism, and with changes in the physical environment such as land abandonment and forest felling. This reveals the countryside as a place where there is as much social deprivation as there is beauty and this, coupled with the presence of an older rural population and marginal location, results in a high degree of social exclusion.

One of the key factors that affects perception of the landscape and seems to account for much of the nostalgic feelings expressed in the findings is the time spent in the countryside as a child. Nearly 82% of the interviewees now living in the towns and cities grew up in, or regularly visited, the countryside as children and this has a marked effect on the idea of what the countryside is or should be, as well as accounting for much of the association with the sense of Latvian identity.

As part of the data analysis, a factor analysis was carried out. The first factor to emerge connected several questions about life in the countryside on which responses (on a five point scale from 'strongly agree' to 'strongly disagree') were used:

> 'I will continue to live in the countryside if more services are available'
> 'I will continue to live in the countryside if there is employment available'
> 'I will continue to live in the countryside'
> 'I would like to bring up my children in the countryside'

SIMON BELL AND ALICIA MONTARZINO

'I will continue to live in the countryside if more services are available'

There is a clear preference among current rural dwellers for continuing to live in the countryside if services are available, suggesting that the lack of services in some areas is a major problem. Services include shops, public transport, schools and medical facilities, postal services and so on. This list is similar to the situation that would be found in other rural locations (see, e.g. Bell, 2003). The pattern varies by age, with more people in the older age groups tending to agree and fewer in the younger groups. This could be related to the preferences of younger people to leave to go to cities or abroad. It also highlights the challenges for the older people who want to continue to live in the countryside and for whom medical and social services are particularly important, as well as transport.

'I will continue to live in the countryside if there is employment available'

The question of employment is a factor affecting whether rural dwellers can continue to live there and whether people currently living in towns would wish to move back to the countryside. Clearly, people need an income. The question is then whether they actually work in the countryside or commute to towns for work, which depends on where they live and the degree of remoteness and distance from potential employers.

For the younger age groups, especially teenagers and young adults under 20 years of age, even the presence of jobs does not seem to make living in the countryside especially attractive when compared with the population aged over 20, although this pattern of response is more pronounced among the urban dwellers than among the rural population.

It seems to be the availability of more services that would make the countryside more attractive for both urban and rural dwellers, rather than the availability of jobs. This may be affected by the improved life of those who live in the countryside – or who would like to return/move there – when services are good, even if they prefer to commute to a town for a better job. This may also be connected with the perception that services are important for reducing feelings of isolation from the rest of society.

'I will continue to live in the countryside'

This might be seen as a way of identifying those who are country people at heart and who are willing to make certain

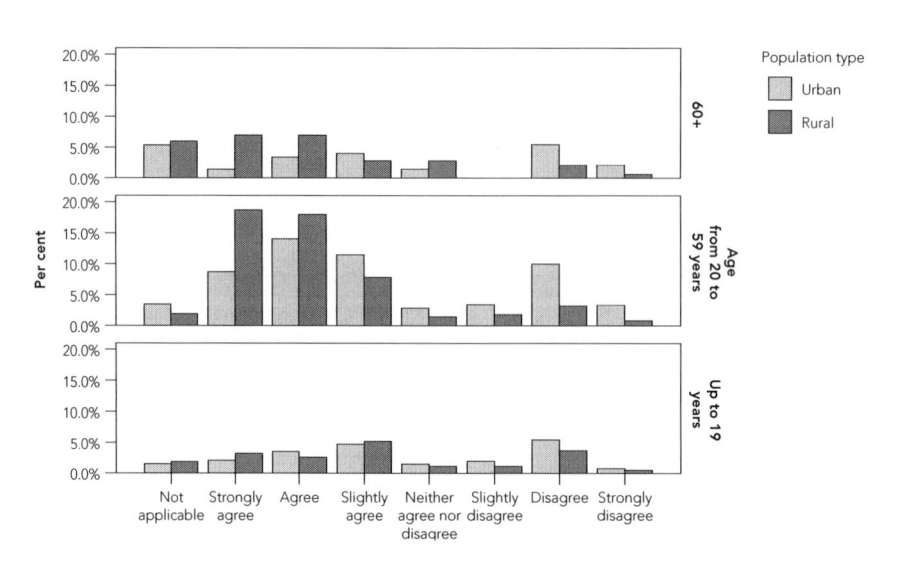

5.7 Graph showing the degree of agreement/ disagreement with the statement 'I will continue to live in the countryside if there is employment available'.

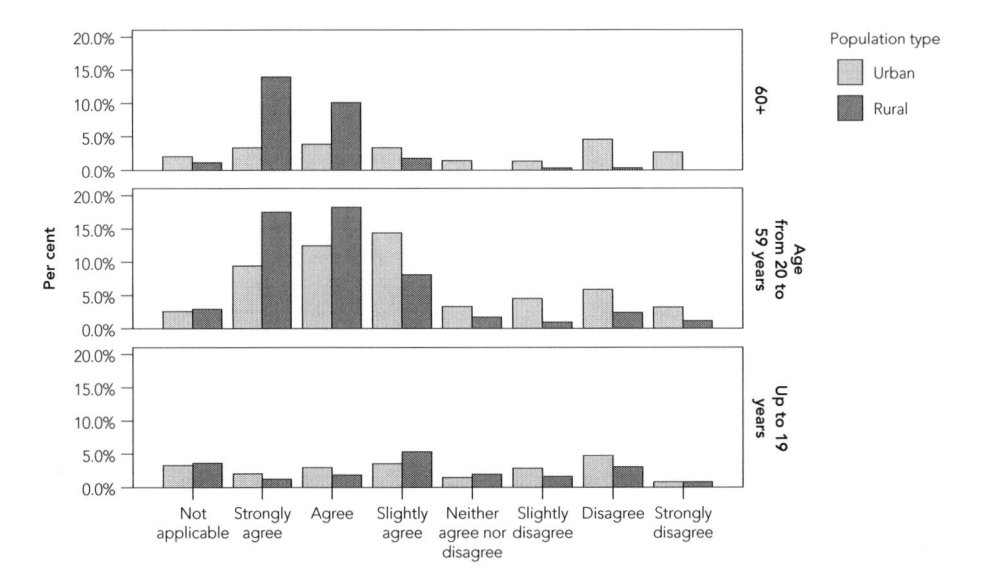

5.8 Graph showing the degree of agreement/disagreement with the statement 'I will continue to live in the countryside'.

sacrifices in material standards of living in order to continue to live in the countryside. There are similar patterns in response to this statement as to the previous ones. The rural dwellers especially tend to agree with the statement quite strongly, as might be expected, but within this population, the younger age groups are far less interested in living in the countryside than the older age groups. The mid-age range sample shows more mixed views, some wanting to stay in the countryside, others wanting to leave. This reflects similar findings by Nikodemus et al. (2005).

'I would like to bring up my children in the countryside'

Those who spent all their childhood in the countryside, regardless of where they live now, show a much higher level of agreement with this statement than those who did not, or who only spent a part of their childhood there.

Discussion

The research has uncovered some key findings that may be common to people in rural areas across Europe. In the two countries discussed the landscape has followed a different trajectory. In Scotland the depopulation of the Highlands took place over 150 years ago and the population is stable with a degree of inward and outward migration, whereas in Latvia the process of rural depopulation that was prevented during the Soviet era (because people were not allowed to leave the *kolkhoz*) is now in full swing. Land ownership patterns are also very different: the large estates in Scotland give a degree of stability to land use, although forest areas have increased through planting, while in Latvia the large estates of pre-independence times and the later 'ownerships' of the Soviet era have reverted to a pattern of many small landowners.

However, in both Scotland and Latvia, people want to continue to live in the countryside but find the lack of services a major obstacle to enjoying an adequate quality of life. The problem of finding a job also affects both locations. In both places there is a strong association with the countryside. In Latvia the countryside is identified with being Latvian and the

SIMON BELL AND ALICIA MONTARZINO

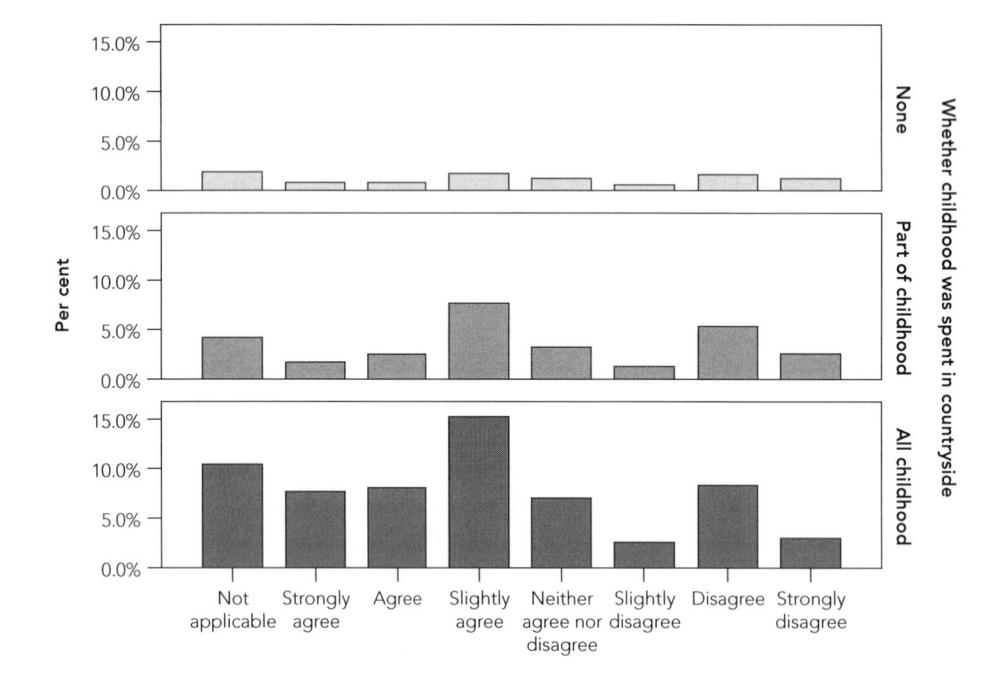

5.9 Graph showing the degree of agreement/disagreement with the statement 'I would like to bring up my children in the countryside'.

special place the countryside has in Latvian culture suggests that the ties to the land are still strong and affect quality of life, as suggested by Vermuri and Costanza (2005). This is perhaps to be expected, since there is a greater proportion of the population still living there or who were brought up there than in many other countries. This is echoed in Scotland in the association of people with Strathdon as an identifiable place, in part accounted for through their ancestral connections, many people being able to trace their family presence there for over 100 years, and by the unique traditions associated with the area, such as the Lonach.

Both projects illustrate the way that aspects of the social environment intersect with that of the physical environment in people's perceptions, attitudes and expectations. This confirms similar findings in other research: the physical environment has a significant role to play in everyday life, but this role is rarely explicit unless a feature of the environment obstructs, prevents or otherwise interferes with a person's objectives (Scott and Canter, 1997). The physical environment provides a setting for the social environment. The relationship between the physical and social environment is transactional – if changes are made to the physical environment, whether people think they are good or bad will depend on the extent to which they affect how people can carry out their jobs and tasks. Thus, as noted in Strathdon, changes through felling the forest or planting more trees, for example, could affect people differently depending on whether they were a logger, landowner or hotel proprietor. Likewise, land abandonment or forest felling will be seen differently depending on the person's relationship with the area, whether they live in a remote lonely place surrounded by scrub, for example, or whether they drive a timber lorry for a living.

However, in Latvia, where the greater volume of data permits more detailed analysis, these associations can also be shown to be strongly related to age. It might be expected that the nature of the relationship will change, as the older generation is replaced by younger ones, for whom this connection with rural place appears to be weaker. Even so, among all age groups there are strongly recognizable, archetypal countryside elements that are associated with making the landscape distinctly Latvian and which appear to be highly valued. The maintenance of the traditional landscape has been identified by Busmanis *et al.* (2001) as dependent on the traditions of the rural life-style and the single farm integrated into the natural environment.

The dichotomy of perceptions highlighted in Latvia by the choice of words recorded by the questionnaire respondents is particularly interesting. The words were selected by the respondents themselves and not chosen from a pre-designed list. Moreover, they were not asked for positive or negative words, only words which they felt were associated with the Latvian countryside. The frequency of the use of the words gives a good idea of the importance of these factors, while the mix of positive and negative terms highlights the mixed feelings of the respondents.

Although, overall, services appear to be more important than jobs for the inhabitants of both areas, there is little doubt that jobs play an important role in whether people continue to live in the countryside, commute to local towns, move to urban areas or even work abroad. The massive increase in travel abroad to find work has been a recent phenomenon that started once Latvia joined the European Union in 2004, after the research findings were collected. This trend bears out the perceptions and attitudes about unemployment and the desire to escape from a marginalized rural existence. Depending on how long this trend for overseas work continues for Latvians, it also has implications for the people left behind. News reports (the only evidence available so far) suggest that in some villages in Latgale, the only people left are the old, the retired and the children (BBC, 2006). Strathdon people also struggle with finding adequately paid jobs in the area itself.

Since there is a close association between positive perceptions of the countryside and having been brought up there, it seems likely that as Latvians become more urbanized, this association may weaken. The Strathdon evidence showed that those living there have strong associations, but the feelings of those who live in cities was not explored. Already in Latvia, since the restitution of land to the former owners, many such owners are not resident on the land and have neglected it, which is one reason so much land is abandoned. If the descendants of older people still living in the landscape do not want to live there, then the property may be sold, may be abandoned or may be used as a holiday house with the land let out to other farmers or left unmanaged. Currently, Latvians tend to spend a good part of the summer in the country and this could continue to maintain the strong cultural association with the countryside as part of the sense of Latvian identity. Some of the more attractive areas may then become significant holiday locations but fail to maintain a year-round viable population, or be able to support an adequate level of services for those who remain living there.

Another developing trend is for people who currently live in flats in large, Soviet-era apartment complexes in the suburbs of Riga to move to single family dwellings on new developments on the edge of the city or in the surrounding countryside. As the economy develops and incomes rise, this trend may continue and the type of gentrification of the countryside that is common in countries such as the United Kingdom, with commuters living there (Spencer, 1997), may become more popular. If transport infrastructure improves, the potential commuting distance from Riga could increase to encompass a significant area of the country, leading to a revitalisation of the wider region. This may help to keep some infrastructure available and help to maintain houses but may not have much effect on land abandonment, the poverty or access to services of older people or, for that matter, the fate of areas outside the range of commuters. This could lead to a two-tier countryside – a gentrified top tier, where well-off commuters live in rural areas within an hour's travelling distance of Riga, side by side with retired or unemployed poorer people, beyond which is a second tier of countryside emptied of all but the older people and others trapped by unemployment or poverty. The pattern of commuting in Scotland shows that the rural areas with the highest number of residents are those within 30 minutes' drive of a town. These areas have higher property prices and schools, shops and post offices remain viable. Strathdon lies outside this zone of influence and suffers accordingly, although incomers live there who prefer to earn a living locally, much like those areas of Latvia ouside the Riga commuting distance.

When comparing Strathdon and Latvia, it is clear that similar issues emerge; in particular, the transactional nature of people's relationship with place. In both places the social environment plays a dominant role in relation to the physical environment, but in both the attractive qualities of the rural landscape and its quietness are powerful attractors for people continuing to live there. While the rate of re-population of some parts of the countryside by urban people is already high in the United Kingdom, and visible in Strathdon, it is only just starting to happen in Latvia.

It is also clear that some people in both Strathdon and Latvia have such strong associations and identify so closely with their local landscape that they are prepared to tolerate a certain level of social and economic disadvantage because they see in it a quality of life, derived from the qualities of the physical environment, that is worth retaining.

References

BBC news, www.news.bbc.co.uk/1/hi/world/europe/4786134.stm (visited 11 December 2006).
Bell, S. (ed.) (2003) *Crossplan: Integrated, Participatory Landscape Planning as a Tool for Rural Development*, Edinburgh: Forestry Commission.
Borg, I. and Shye, S. (1995) *Facet Theory: Form and Content*. London: Sage.
Breakwell, G.M. (1990) *Interviewing*, Exeter: British Psychological Society.
Breakwell, G.M. and Canter, D.V. (1993) *Empirical Approaches to Social Representations*. Oxford: Clarendon Press.
Busmanis, P., Zobena, A., Grinfelde, I. and Dzalbe, I. (2001) 'Privatisation and soil in Latvia – land abandonment', paper presented at seminar on Sustainable Agriculture in Central and Eastern European Countries: The Environmental Effects of Transition and Needs for Change, Nitra, Slovakia.
Canter, D. (1977) *The Psychology of Place*. London: Architectural Press.
Casely, G. [2000] see http://users.tinyonline.co.uk/amchardy/McHardy/LonachwhatisLonach.htm (visited on 11 December 2006).
Commission for Rural Communities (2006) *Rural Disadvantage: Priorities for Action*. Cheltenham: Commission for Rural Communities.
Deller, S., Tsai, T.-H., Marcouiller, D.W. and English, D.B.K. (2001) 'The role of amenities and quality of life in rural economic growth', in *American Journal of Agricultural Economics*, vol. 83 (2), pp. 352–365.
Earthtrends, www.earthtrends.wri.org (visited on 11 December 2006).
Eurostat, www.epp.eurostat.ec.europa.eu (visited on 11 December 2006).
Grave, Z.L. and Lūse, M. (1990) 'Designing and practice of rural settlement in Latvia', in *Proceedings of the Latvian Academy of Sciences*, vol. 7, pp. 76–85 (in Russian).
Kelly, G. (1955) *Principles of Personal Construct Psychology*, vols I–II. New York: Norton.
Kinsella, J., Wilson, S., de Jong, F. and Renting, H. (2000) 'Pluriactivity as a livelihood strategy in Irish farm households and its role in rural development', in *Sociologia Ruralis*, vol. 40 (4), pp. 481–496 (16).
Korpela, K.M. (1992) 'Adolescents' favourite places and environmental self-regulation', in *Journal of Environmental Psychology*, vol. 12, pp. 249–258.
Lūse, M. and Jakobsone, A. (1990) 'Development of the idea of country villages', in *Proceedings of the Latvian Academy of Sciences*, vol. 7, pp. 87–97 (in Russian).
Madgeley, J. and Bradshaw, R. (2006) *Should I Go or Should I Stay?* Newcastle upon Tyne: IPPR.
Melluma, A. (1994) 'Metamorphoses of Latvian landscapes during fifty years of Soviet rule', in *Geojournal*, vol. 33 (1), pp. 55–62.
Nikodemus, O., Bell, S., Grine, I. and Liepiņš, I. (2005) 'The impact of economic, social and political factors on the landscape structure of the Vidzeme uplands in Latvia', in *Landscape and Urban Planning*, vol. 70, pp. 57–67.
Prebble, J. (1963) *The Highland Clearances*. London: Penguin.
Robson, C. (1993) *Real World Research*. Oxford: Blackwell.
Rural Poverty and Exclusion Working Group (2001) *Poverty and Social Exclusion in Rural Scotland*. Edinburgh: Rural Poverty and Exclusion Working Group.
Scharf, T. and Bartlett, B. (2006) *Rural Disadvantage. Quality of life and disadvantage amongst older people: a pilot study*. Cheltenham: Commission for Rural Communities.
Scott, M.J. and Canter, D.V. (1997) 'Picture or place? A multiple sorting study of landscape', in *Journal of Environmental Psychology*, vol. 17, pp. 263–281.
Scottish Executive (2005) *Rural Scotland, key facts*. Edinburgh: Scottish Executive.
Shye, S., Elizur, D. and Hoffman, M. (1994) 'Introduction to facet theory; content design and intrinsic data analysis', in *Behavioural Research*, Thousand Oaks, CA: Sage.
Spencer, D. (1997) 'Counter-urbanisation and rural depopulation revisited: landowners, planners and rural development', in *Journal of Rural Studies*, vol. 13 (1), pp. 75–92.
Stockdale, A. (in press) 'Migration: prerequisite for rural economic regeneration?', in *Journal of Rural Studies*.

Twigger-Ross, C.L. and Uzzell, D.L. (1996) 'Place and identity processes', in *Journal of Environmental Psychology*, vol. 16, pp. 205–220.

van der Ploeg, J.D. *et al.* (2000) 'Rural development: from practices and policies towards theory', in *Sociologica Ruralis*, vol. 40(4), pp. 391–408.

Vemuri, A. and Costanza, R. (2005) 'The role of human, social, built, and natural capital in explaining life satisfaction at the country level: Toward a National Well-Being Index', in *Ecological Economics*, vol. 58, pp. 119–133.

Ward Thompson, C. and Scott Myers, M. (2003) 'Interviews and questionnaires', in S. Bell (ed.) *Crossplan: Integrated, Participatory Landscape Planning as a Tool for Rural Development*. Edinburgh: Forestry Commission.

Westhoek H.J., van den Berg, M. and Bakkes, J.A. (2006) 'Scenario development to explore the future of Europe's rural areas', in *Agriculture, Ecosystems and Environment*, vol. 114, pp. 7–20.

CHAPTER 6

Mapping youth spaces in the public realm

Identity, space and social exclusion

Penny Travlou

Introduction

Young people have needs in terms of access to environments that support their healthy development as individuals and members of society; engagement with the public domain is an important part of this development and public space offers a key environment for teenagers/young people. However, young people are by and large unwelcome in public space and suffer from exclusionary treatment by other members of society. This chapter explores teenagers' engagement with public space in the city and argues for a more open and inclusionary approach to understanding young people's needs. Although the focus is on young people's use and experiences of Edinburgh city centre in Scotland, aspects of the study have been shown to have an international relevance (Travlou et al., forthcoming).

For many young people, public space is a stage for performance and contest, where a developing sense of self-identity is tested out in relation to their peers and other members of society (Travlou, 2003; Ward Thompson et al., 2004). Young people – both boys and girls – spend a large amount of their free time outside their homes, 'hanging out' with friends on the streets and in other public domains. They often use public or quasi-public spaces to hang out as those places offer them more autonomy, anonymity and freedom from parental supervision. According to Lieberg, 'teenagers have no obvious right to spaces of their own. They often have nowhere to go except public spaces, where they often come into conflict with other groups' (1995: 720). Hanging out and about on streets, in public parks, shopping malls, urban woodlands and city centres in general, renders teenagers visible and their visibility places their behaviour under scrutiny (Bell et al., 2003). The response from other groups in society is often negative; young people are predominantly perceived as a problem, responsible for crime in public space. While younger children are seen as too innocent and vulnerable to dangers in public space, older children are often confronted as the primary culprits of disturbance.

> From being innocent and vulnerable 'angels', victims of circumstance, in need of care and protection, children in trouble have been systematically reconstructed and

(re)presented in the late 1990's as 'demons', the knowing perpetrators of malevolent and evil acts.

(Matthews et al., 1999: 1713)

The evidence of this attitude in British society is revealed in the development and use of the Anti-Social Behaviour Order (ASBO), a civil order made against a person who has engaged in conduct which 'caused or was likely to cause alarm, harassment, or distress to one or more persons not of the same household as him or herself' (Office of Public Sector Information, 1998; Home Office, 2004). Anti-Social Behaviour Orders, first introduced in 1998, have become a tool used increasingly to tackle youth crime and disorderly behaviour. Between 2001 and 2004, 46% of the 3,500 ASBOs issued were served on minors, some as young as ten years old, but the majority aged between 14 and 16 (House of Commons, 2005). Although ASBOs have been welcomed by many community groups, they have also been criticised for dealing with problems, not causes, and for perhaps exacerbating situations which generate anti-social behaviour, for example where young people find themselves constrained into physical and social environments, often indoors, which add to their problems and frustrations.

Young people's demonization and subjection to public scrutiny and state controls is only one aspect of the problematic position of young people in the public realm. Besides being marginalised and excluded from *adults'* public space, young people also have to confront the hostility of other teenage groups who want to control the local areas where they 'hang out' (Matthews et al., 1998; Woolley et al., 1999; Nairn et al., 2000; Percy-Smith and Matthews, 2001; Travlou, 2004).

Although the contest for 'ownership' of places is central to the ways young people use and experience public space, it is under-researched. Most studies on young people's perception of public space focus either on social exclusion or on crime and vandalism (e.g. Cahill, 1990; White, 1993; Matthews, 1995; Valentine, 1996a, 1996b; Woolley et al., 1999). By failing to take into account the broader canvas of young people's 'ways of seeing', teenagers are treated as an outsider group. Matthews (1995) suggests that there is a need to investigate the environment as young people understand it, as only in this way can they become fully integrated users of large-scale places. The research described in this chapter goes some way towards redressing the balance by giving young people a voice to express their environmental values and perceptions. It uses the technique of place mapping (Travlou et al., forthcoming) to illustrate young people's movement in the city: places frequented by different youth groups, routes taken or avoided, safe havens and spots that could expose them to harassment. It places emphasis on the ways in which teenagers influence and form their own culture within small groups – microculture – and how this manifests itself in terms of their environmental behaviour. It illustrates how young people's understanding of public space is very often unconventional and beyond adults' expectations.

Background to the study

The research described here is based on a three-year study concerning young people's attitudes to and uses and experiences of urban public space in Edinburgh. This project is part of an international collaboration between three centres, in Scotland (OPENspace research centre at Edinburgh College of Art) and the United States (University of California, Davis and Cornell University, New York State). The focus of the investigation is teenagers and public places: it explores teenagers' perceptions of public space and the meanings behind these perceptions. A number of locations (Sacramento in California, New York City and Ithaca in New York State and Edinburgh in Scotland) were used for case studies. In the course of this multi-site, case study approach, a range of methodologies was tried and refined by each research team and initial results offer insights into common patterns of teenage experience.

The research focused on young people aged 12 or over, that is, at an age when a certain amount of independence is usually allowed, and up to the age of 18. One challenge in this study has been the finding that much of the material on children, young people and the use of open spaces relates to children aged 11 years or younger. Although there is clearly a continuum of experience from early childhood to adolescence and teenage years, we have tried to maintain a focus on the older age group, as defined. We recognize the importance of earlier childhood experience in relation to teenagers' use of the urban environment, but the older group's needs appear to have been less well served to date.

The main question which this project addressed is whether teenagers are a socially excluded group with regard to public open space in Edinburgh. To answer this question, we looked at how young people engage with public space in the urban environment, across various types of community. We investigated the ways in which young people and teenagers (12- to 18-year-olds) perceive the built and physical environment in relation to their use of different places and explored the symbolic significance this use has for them in the context of geographically and culturally disparate communities. In particular, we examined young people's experience of Edinburgh city centre, focusing on public places such as streets, parks and squares beyond the home, school and playground environment. We asked teenagers to describe the places they like to go in their city, examining young people's engagement with their environment according to the components of place: physical attributes of the environment, young people's behaviours and their conceptualizations of place (Canter, 1977).

The emphasis has been on the ways in which young people shape their own culture within small groups and how this manifests itself in terms of their environmental behaviour – how they create their own *microgeographies* within their communities (Matthews *et al.*, 1998). The project also aims to explore the physical nature of the outdoor public realm and the patterns and structures of space and societal context which permit or exclude teenagers' use.

Methodology

This study follows a qualitative methodological approach, comprising focus group discussions and place mapping exercises with school pupils as well as site observations of selected locations in Edinburgh city centre. A detailed review of current literature on young people's use of public space (Travlou, 2003) highlighted gaps in understanding and allowed us to refine our research approach.

The study took place in Edinburgh, the historic capital of Scotland. Situated in east-central Scotland, Edinburgh is the country's administrative, cultural, educational and service-industry hub. It is the second most populous city in Scotland after Glasgow. According to the last Scottish Population Census in 2001, Edinburgh's total population for that year was 430,082, of which 16.11% were children and young people under 16 (Scotland's Census Results Online, 2001).

Ten centrally located Edinburgh schools were invited to participate in the research; Edinburgh is atypical of Scotland in having 24% of pupils in private education, so private schools were included in the invitation. The four schools finally chosen were all co-educational and gave access to young people across a wide range of ethnic and socio-economic backgrounds, partly related to school type (publicly or privately funded) and to location. There were three local authority funded schools, one centrally located and two in more peripheral locations in the city, and a privately funded school. Fifteen focus group discussions were carried out with a total of 150 pupils from two age groups, 12–14- and 17–18-year-olds. The interviews were based on a series of prepared questions relating to physical and social attributes of public space, young people's activities in it, and their perceptions about it. Taking a user-led approach, we started off by asking participants for their perceptions of public space. Sub-questions included: where do teenagers go out for social reasons (as opposed to school requirement, etc.)? When do they go (time of day, day of week, season(s), etc.)? How far are these places from home and how do they get there and back (mode of transport)? What do they do in outdoor spaces? Whom do they go with and whom do they meet up with? What are the things they like about the places they go, what do they dislike? Where do they go when they feel happy and where do they go when they feel angry or sad? What would their ideal outdoor place be like?

Six of the focus groups carried out towards the end of the data-gathering period (2004) involved an innovative approach – 'place-mapping' – as an additional method (Travlou *et al.*, forthcoming). We became aware that focus group discussions alone were not necessarily revealing the full spectrum of young people's experience(s) in and around Edinburgh, particularly in relation to the physical environment. Place mapping, as utilized in our research, was developed as a technique to locate and elicit comment about places that play a significant role for youth, both positively and negatively, and to provide a common point for discussion among a group of teenagers. The method offers the researcher an opportunity to go beyond the map in exploring young people's spatial experiences with regard to inclusion and exclusion from public space. Place mapping contributed to an understanding of young people's engagement as a social group with everyday spaces in the urban nexus. The maps became the focus of a discussion about these places that went beyond their physical location and facilitated a debate about the negotiated meanings of the places, as well as their physical qualities and the behaviours associated with them.

Main findings

The key findings from the study showed that teenagers value public places first and foremost for the sense of spatial autonomy they allow and for the opportunities for social interaction they may offer. The research data – transcripts from the focus groups and place mapping discussions – highlight how the opportunity to be with friends when outside home and school, away from adult supervision, seems to be more important to young people than the physical character of a place. The physical characteristics of a place were secondary in their choice of favourite places. When asked what they valued about their favourite places to go to, the teenagers in this study most often mentioned the social characteristics of such places, that is, a place where they could be with friends or meet new people of their age. Very often, they mentioned that the places they like to go to would have been boring without the company of their friends.

> 'If I'm on my own, there's not much point [to go to town] because it's quite boring and you just walk around and you don't have anything to do.'
> (16-year-old boy)

> Girl 1: 'I don't like going there [to Edinburgh city centre] on my own, 'cause it would be boring.'
> Girl 2: 'Yeah, you wouldn't have anyone to talk to.'
> (12-year-old girls)

Furthermore, being away from home, school or/and local neighbourhood – what Matthews and Limb call the 'fourth environment' (2000) – blurs our respondents' distinction between indoors and outdoors. When asked to describe their favourite outdoor places, their responses included indoor environments such as shops and shopping malls. Evidently, for our respondents, outdoors often meant outwith the fourth environment.

> 'It doesn't matter where you go but with whom you go.'
> (16-year-old girl)

> 'Somewhere where you could go inside or outside . . . that's why it's like Princes Street because you could go out and then if you want you could go in one of the shops and . . . then somewhere where they would be people of your age.'
> (15-year-old boy)

For the majority of teenagers, their favourite spots for hanging out are commercial streets or centres, shopping malls and multi-complex cinemas. Local shopping areas also provided places of social contact, where young people could meet and relax. In essence, these are adults' places, designed to meet other requirements, which have been appropriated by teenagers. They are places that offer them the possibility of social interaction, freedom and anonymity, safety and accessibility, a variety of amenities and opportunities for consumption.

As teenagers get older and become freer to travel and use the city centre, they also become much more economically significant to those areas. They have a significant aggregate purchasing power, as their personal focus shifts from their local neighbourhood area to the central commercial and leisure areas of Edinburgh city centre. Their exercise of choice is both socially and commercially important in the fields of entertainment, sport and purchase of clothes and food.

Princes Street – Edinburgh's high street – was nominated by most of our interviewees as their favourite place in the city, having all the above characteristics appealing to young people. They described it first as a safe place where they could socialize with their friends and, second, as a place of consumption with a variety of amenities for shopping and entertainment.

MAPPING YOUTH SPACES IN THE PUBLIC REALM

'My favourite place is Princes Street. There are places to drink, and places to eat and places to shop, all sorts of entertainment.'

(16-year-old girl)

'Shopping is like socialising, 'cause you go there [Princes Street] with your friends, you don't go alone.'

(14-year-old girl)

However, these same places were described by some of the participants as being sometimes unwelcome and unsafe. The presence of groups perceived as dangerous in the same places that teenagers frequent reflects the multi-use character of the city. Our participants referred to the presence of dangerous groups of adults as well as to other teenage groups. What distinguishes their experience of 'hanging out' is their use of different places according to the youth culture to which they belong. Each group appropriates some space, creating distinct places fashioned by their microculture (Travlou, 2004). In our research, the respondents mapped the city according to locations frequented by specific teenage groups to which they gave labels, for example 'chavs' (a term, often used in a derogatory way, to denote a youth subculture associated with young people from council housing estates in areas of deprivation), 'goths', 'skateboarders', and so on. Not surprisingly, some of the respondents identified themselves as members of these subcultural groups. One could proceed much further with identification and description of teenage subcultures in Edinburgh, identified by musical preferences, dress code or engagement with extreme sports, but these three groups were the ones most consistently singled out by teenagers themselves during the course of our research. It is clear that there are many different kinds of young people hanging out in the public spaces of Edinburgh, each with their own mode of appropriating space and constructing identity. It is misleading, therefore, to treat young people as a homogeneous group with the same needs, expectations and experiences of public space.

'Goth' kids, also known as 'scary' kids, and skateboarders are two subcultural groups whose presence in Edinburgh public space is associated with considerable controversy. Due to the style of their use of public space, these groups are often met with hostility not only by adults but by other teenagers too.

Both groups have adopted particular places in the city centre as spatial hubs, where they can be very visible while at the same time away from other, potentially hostile teenage groups. These places become a safe space where goths and skateboarders can gather, to affirm their sense of difference and celebrate their feelings of belonging to a group. Their difference, however, is an identifying tag that can attract the dislike, even the hostility, of other teenagers. Throughout our focus groups, a great number of participants expressed their dislike for those youth groups, arguing that their presence in particular central locations in the city make them avoid these areas. They felt that those places (i.e. streets, parks) belonged to these teenagers who, through their attitude, dress and behaviour, had appropriated the place to other participants' exclusion (see Figure 6.1).

'Cockburn Street is awful; it's full of goths. It doesn't bother me what they do; it's just the idea that they are there and they don't go anywhere else. They stand there in big groups and you have to make your way through the group. The problem is that Cockburn Street has become THEIR STREET. No one else could go there. They think that they own the street.'

(15-year-old boy)

For teenagers who belong to such youth subcultures, on the other hand, the street is their playground, their meeting place: they feel safe there, away from other teenagers who may provoke them.

'It's safe here [Cockburn Street] – a laugh, with no 'schemies' to provoke us. We're unfairly labelled.'

(15-year-old girl)

One teenage group that faces exclusion from public space as well as hostility from other young people are skateboarders (Stratford, 2002). This group of 'active' teenagers has caused such a major impact with their presence in public space in Edinburgh (i.e. parks, streets, public steps) that the public has become polarized into those who tolerate skateboarders and those who dislike them and want them out of streets and squares (see also Jones and Graves, 2000). Young people's attitudes towards skateboarders reflect a more general belief

6.1 Young people – 'goths'.

Credit: Image used by kind permission © Fin Macrae 2003 (www.finmacrae.com)

MAPPING YOUTH SPACES IN THE PUBLIC REALM

that their strong presence in certain public places prevents many non-skateboarders from using them too.

'I don't like the skateboarders; I don't like them running around with their stupid wooden boards and causing such a noise and havoc.'

(14-year-old boy)

'I think the best is to give them their own place, a skatepark where they could do whatever they want and leave the rest of us alone.'

(16-year-old boy)

Bristo Square, a central square in the university area, is a good example of skateboarding space; skaters are the most conspicuous of its teenage users. The number of skateboarders in the square is on the increase as there is as yet no specially designed or dedicated skatepark in the city (see Figures 6.2 and 6.3). The intention to build a skatepark was announced two years ago by the local authority, in an effort to resolve spatial antagonism

6.2 Skateboarding in Bristo Square, Edinburgh.

6.3 Skateboarders as a youth subculture, Bristo Square, Edinburgh.

between skateboarders and other users of public places in Edinburgh. Disputes about the location of the skatepark, however, have prevented any progress with implementing the project. Skateboarders themselves, when asked about their experience of using public places in Edinburgh and antagonism with other users – among them other young people – argued that they also need places where they can feel safe, as well as free of any adult supervision and in control of space. They felt that others have misunderstood them and thought of them only as a nuisance.

> 'I like skateboarding but there are few places to skateboard because we're kicked out all the time from places we like to skateboard.'
>
> (14-year-old boy, Bristo Square)

> 'They said that they would build a skatepark but I don't think that's happening . . . Then apparently there's also a skatepark being built near the airport; it's the first one ever in Edinburgh . . . Yeah it's a bit far away but the second nearest one at the moment is near Livingston and after that is Dundee and then that's it!'
>
> (16-year-old boy)

Skateboarders also complained that ownership and control of their 'own' spaces is not uncontested by other teenage groups.

> 'In Portobello, there's a big kind of indoor skatepark but there're lots of chavs around there . . . I used to go there when I was quite young . . . like four or five years ago. The last time I went, I got nervous 'cos these big people, they'd kind of bully you and force you to keep going and try stuff you couldn't do, sometimes they'd make you not wanna go even if you quite want to do it.'
>
> (15-year-old boy)

Discussion of main findings

Place maps created by Edinburgh youth revealed young people's experience of the city centre as both an 'urban stage' and an arena of contested places. Their experience oscillated between favourite and least favourite, safe and dangerous places. The presence of other teenage groups in these central places is seen by many teenagers – particularly the younger age groups – as a drawback. This is because the spatial co-existence of different teenage groups may result in hostile and aggressive behaviour as a way to draw spatial boundaries. For instance, Matthews et al. (1998: 196) discovered that:

> 'Hassle' from other, often older, 'kids' and fear of assault among the girls, and fear of attack and fear of fights among the boys, kept these teenagers to tightly defined areas, where they felt 'safe' and free to do what they wanted.

Research into young people's use of public space showed that young people commonly fear being in their local areas while other teenage groups are present (James, 1986; McLaughlin, 1993; Percy-Smith and Matthews, 2001). As an expression of young people's contesting microgeographies, neighbourhood bullying is understood as the way:

> Different groups use particular places, such as the neighbourhood, to play out identity struggles between self and others . . . in terms of shared interests, behaviours and circumstances which often give rise to multilayered micogeographies co-existing in the same location.
>
> (Percy-Smith and Matthews 2001: 52–53)

Some theorists even talk about a kind of 'spatial apartheid' imposed as teenage groups draw social boundaries and exercise control on the landscape (Tucker, 2003: 121). Inevitably, the outcome of these struggles influences young people's spatial behaviour. This type of territorial behaviour could be defined as:

> The behaviour of an individual (or group) claiming control over a particular area. This behaviour relates mainly to the area itself and includes the definition and marking of the area and its defence from intruders of the individual's own kind.
>
> (Sebba and Churchman, 1983: 191)

Drawing on the above discussion, the focus group interviews revealed that issues such as fear about one's personal safety and the presence of dangerous teenage groups are of main concern for young people. Quite often they referred to incidences of verbal and physical harassment by other teenagers. In all of our focus groups, our young participants – regardless of type of school – either talked about their fear of being harassed or described incidents of personal harassment by other teenage groups:

'Yeah, there're all these gangs and they meet and do things. They have all different names but now there's more patrol and police so it's a bit better but still it is not brilliant. They have all different weapons like knives and razor-blades. They are from my age to about twenty sometimes. It was once that I was going back home and that guy came to me and said things and then his friends came too and they were all around me but mainly they were saying things.'

(16-year-old boy)

'At the park they [teenage gangs] all hang around on the swings and whenever you go to the park they tell you to go away, yeah, and if you're like walking past they'll chuck the swings at you and see if they can hit you and stuff.'

(14-year-old girl)

Some of the respondents tried to avoid certain areas in the city where they knew that they would meet other teenage groups and could possibly get in trouble. One 15-year-old boy described how worried he was for his personal safety:

'I wouldn't go to any area where there's a gang!'

Some pupils described their fear of using public transport to go to the city centre, afraid that teenage gangs would attack them, verbally or physically, at the bus stop or on buses. They said that the violence was sometimes related to school rivalries and uniforms.

Generally speaking, when bullying occurs, the dominant group controls or exploits the encounter space through the imposition of their own sets of rules and values to the detriment of the other, undermining the other's developing autonomy and capacity for agency (Percy-Smith and Matthews, 2001). From this perspective, encounter spaces are not necessarily set domains or strongly demarcated territories, but fluid spaces of interpersonal interaction. For the unfortunate young people who become victims of bullying, local environments can become tyrannical spaces, defined in terms of 'no-go areas', danger and threat (Percy-Smith and Matthews, 2001).

The maps of teenagers' least favourite places portrayed the city centre as being a contested space consisting of dangerous places, of boundaries and territories, a 'turf map' where different groups of young people claimed ownership of different places, from a single street to a corner shop (see Moore, 1986: 32). The primary organization of space was a division of places into those which young people considered safe and affirming, and those which they viewed as negative or from which they felt excluded. An important way in which young people signalled their involvement with certain kinds of place and identity was effected partially through the divisions and exclusions that young people themselves created. This double-sided process of inclusion and exclusion appeared to be crucial in orienting the young people's sense of self.

Conclusion

This chapter has shown that teenagers' complex personal geographies reflect, and are shaped by, not only tensions between adults and youth in public space but also by youth–youth relations (Tucker, 2003). The technique of place-mapping allied to the focus groups used in this research was particularly effective in revealing the characteristics of urban places as understood by young people.

Tensions between different groups of young people over the use and control of the same places in Edinburgh city centre demonstrate that youth are highly heterogeneous and diverse. Social groupings and spatial antagonism in public space (territorial demands, bullying and friction between social groupings) are the spatial expression of many distinct microcultures. Young people, therefore, cannot be treated as a unified and homogeneous age group. If we are to understand their use and experience of urban space, microcultural differences and distinct preferences should be understood and taken into account. Our research in Edinburgh reveals that there are many different teenage mappings of the city corresponding to the diversity of experiences by various youth groups.

In view of the spatial antagonism between teenage groups, what would the ideal teenage space be like? Turf wars notwithstanding, when asked about their ideal place most Edinburgh teenagers tend to describe it as a hub of co-existence, a teenage central locus specifically made for them with opportunities for all different youth groups. As a girl of 14 years old suggested:

> 'A place for all sorts of different people, even for goths and skaters, all in one place with lots of different things to do.'

This allows us to put teenage contest over public space into perspective. Despite its apparent intensity and the feelings of rivalry and fear it often generates, spatial co-existence of different teenagers is still welcome by them, as an opportunity to interact with others, shape personal and social identities, engage with a world larger than their school and immediate neighbourhood, to be, in other words, citizens. Interesting places are interesting to all – and this adds to their attractiveness. In conclusion, if we want to design spaces for young people, we should think in terms of providing places for social integration and interaction, safety and free movement, accessibility, and variety of activities and amenities.

References

Bell, S., Ward Thompson, C. and Travlou, S.P. (2003) 'Contested views of freedom and control: Children, teenagers and urban fringe woodlands in Central Scotland', in *Urban Forestry and Urban Greening*, vol. 2, pp. 87–100.

Cahill, S. (1990) 'Childhood and public life: Reaffirming biographical divisions', in *Social Problems*, vol. 37, pp. 390–402.

Canter, D. (1977) *The Psychology of Place*. London: Architectural Press.

Home Office (2004) *Perceptions and Experiences of Antisocial Behaviour*. Findings 252. Available at www.homeoffice.gov.uk/rds/pdfs04/r252.pdf.

House of Commons Hansard Written Ministerial Statements for 29 June 2005. 'Antisocial Behaviour Orders'. Available at www.publications.parliament.uk/pa/cm200506/cmhansrd/vo050629/wmstext/50629m02.htm

James, A. (1986) 'Learning to belong: The boundaries of adolescence', in A. Cohen (ed.) *Symbolising Boundaries*. Manchester: Manchester University Press.

Jones, S. and Graves, A. (2000) 'Power play in public space: Skateboard parks as battlegrounds, gifts, and expressions of self', in *Landscape Journal*, vol. 19, pp. 136–148.

Lieberg, M. (1995) 'Teenagers and public space', in *Communication Research*, vol. 22 (6), pp. 720–740.

McLaughlin, M.W. (1993) 'Embedded identities', in S.B. Heath and M.W. McLaughlin (eds) *Identity and Inner City Youth: Beyond Ethnicity and Gender*. New York: Teachers College Press.

Matthews, H. (1995) 'Living on the edge: Children as 'outsiders'', in *Tijdschrift voor Economische en Sociale Geografie*, vol. 86 (5), pp. 456–466.

Matthews, H., Limb, M. and Percy-Smith, B. (1998) 'Changing Worlds: The microgeographies of Young Teenagers', in *Tijdschrift voor Economische en Sociale Geografie*, vol. 89(2), pp. 193–202.

Matthews, H., Limb, M. and Taylor, M. (1999) 'Reclaiming the street: The discourse of curfew', in *Environment and Planning A*, vol. 31, pp. 1713–1730.

Matthews, H. and Limb, M. (2000) 'Exploring the 'fourth environment': Young people's use of place and views of their environment' *Children 5–16 Research Briefing*, Number 9. Available at www.hull.ac.uk/children5to16programme/briefings/matthews.pdf.

Matthews, H., Taylor, M., Sherwood, K., Tucker, F. and Limb, M. (2000) 'Growing up in the countryside: Children and the rural idyll', in *Journal of Rural Studies*, vol. 16, pp. 141–153.

Moore, R.C. (1986) *Childhood's Domain: Play and Place in Child Development*. Berkeley, CA: MIG Communications.

Nairn, K., McCormack, J. and Liepins, R. (2000) 'Having a place or not? Young people's experiences of rural and urban environments', in the proceedings of the *Nordic Youth Research Information Symposium – NYRIS 7*, 7–10 June, Helsinki, Finland.

Office of Public Sector Information (1998) *Crime and Disorder Act 1998: Chapter 37*. Available at www.opsi.gov.uk/acts/acts1998/98037—c.htm.

Percy-Smith, B. and Matthews, H. (2001) 'Tyrannical spaces: Young people, bullying and urban neighbourhoods', in *Local Environment*, vol. 6 (1), pp. 49–63.

Sebba, R. and Churchman, A. (1983) 'Territories and territoriality in the home', in *Environment and Behavior*, vol. 15 (2), pp. 191–210.

Stratford, E. (2002) 'On the edge: A tale of skaters and urban governance', in *Social and Cultural Geography*, vol. 3 (2), pp. 193–206.

Travlou, P.S. (2003) *Teenagers and Public Space: A Literature Review*. Edinburgh: OPENspace. Available at www.openspace.eca.ac.uk/litteenagers.htm.

Travlou, P.S. (2004) 'A teenager's survivor guide to public spaces in Edinburgh: Mapping teenage microgeographies', in *The proceedings of the international conference 'Open Space – People Space', Edinburgh*. Available at www.openspace.eca.ac.uk/conference/conference.htm.

Travlou, P.S, Owens, P.E., Maxwell, L. and Ward Thompson, C. (forthcoming) 'Environment, identity and experience: Place-mapping as a method to understand teenagers' engagement with public places', in *Children's Geographies*.

Tucker, F. (2003) 'Sameness or difference? Exploring girls' use of recreational spaces', in *Children's Geographies*, vol. 1 (1), pp. 111–124.

Valentine, G. (1996a) 'Angels and devils: Moral landscapes of childhood', in *Environment and Planning D: Society and Space*, vol. 14, pp. 581–599.

Valentine, G. (1996b) 'Children should be seen and not heard: The production and transgression of adults' public space', in *Human Geography*, vol. 17 (3), pp. 205–220.

Valentine, G. and McKendrick, J. (1997) 'Children's outdoor play: Exploring parental concerns about children's safety and the changing nature of childhood', in *Geoforum*, vol. 28 (2), pp. 219–235.

Ward Thompson, C., Aspinall, P., Bell, S., Findlay, C., Wherrett, J. and Travlou, P. (2004) *Open Space and Social Inclusion: Local Woodland Use in Central Scotland*. Edinburgh: Forestry Commission.

White, R. (1993) 'Youth and the conflict over urban space', in *Children's Environments*, vol. 10 (1), pp. 85–93.

Woolley, H., Dunn, J., Spencer, C., Short, T. and Rowley, G. (1999) 'Children describe their experience of the city centre: A qualitative study of the fears and concerns which may limit their full participation', in *Landscape Research*, vol. 24 (3), pp. 287–301.

Part 3

Design issues: where are the design challenges and what does inclusive design mean in practice?

What makes a park inclusive and universally designed?

A multi-method approach

Robin C. Moore and Nilda G. Cosco

Introduction

Social inclusion has been the subject of recent initiatives in the United Kingdom and Canada driven by continuing issues of social exclusion of minority ethnic groups, low-income families, people with disabilities, children, youth and elders from mainstream contemporary society. Particularly in the Canadian view, the physical environment and public domain of cities and urban neighbourhoods, including parks, are viewed as critical areas of modern life and, therefore, spaces for social inclusion. As Drache (2001: 8–9) states,

> Environmental inclusion in all cities has to be thought of as the capacity of the physical environment to facilitate and promote sustainable human development. . . . How is the city to become a more inclusive habitat without a process of inclusion anchored in the public domain?

The question turns on the role of urban landscape design in achieving this anchor and challenges designers to provide high quality public spaces that offer more than a merely pleasing physical environment. The question is what tools do park designers need to create such recreational environments that would support social inclusion? This chapter describes a multi-method approach to assess social inclusion in a universally designed park to understand the environment/behaviour dynamics. The approach may be useful to planners and designers wanting to provide successful park environments for all.

The concept of social inclusion goes hand in hand with that of universal design. US architect the late Ron Mace is credited with developing the concept, which he defined as 'the design of products and environments to be usable by all people, to the greatest extent possible, without the need for adaptation or specialized design' (Ostroff, 2001). He saw universal design as an inclusive concept beyond the 'accessible design' of buildings that would accommodate all human needs, including those of people with disabilities (himself a wheelchair user). Development of the term had its beginnings at an expert seminar (co-sponsored by the US National Endowment for the Arts, NEA), including Mace, which reinforced the notion of 'design for all people' as the umbrella concept under which 'accessible design' (the term 'universal design' had not been invented yet) should fit (Ostroff and Iacofano, 1982). This direction in the US discussion on universal design was further advanced in 2003 by a gathering (again sponsored by the US NEA) of US universal design experts who emphasized 'an overarching need for more research in a wide array of critical topic areas . . . types of populations, products, environments and systems [including] . . . urban design and outdoor recreation' (NEA, 2003: 1). Park evaluations were specifically listed in this context. Across the Atlantic, the current UK term closest to universal design is 'inclusive design'. Aligned with the requirements of the Disability Discrimination Act (DDA, Department for Education and Employment, 1995), it has an explicit focus on disability, especially as associated with ageing – the new reality of increasing longevity.

At the young end of the age spectrum, a case can be made for including the general population of children within the purview of universal design because of their vulnerability and developmental needs (Moore et al., 1992). A small proportion of children live with some type of special need (physical, mental or sensory impairment) that requires special environmental modifications, but children as a whole have special needs defined by levels of maturity and skill limitations. Children are also individuals in the process of learning about the world around them. Richer environments – socially, culturally and physically – enhance and extend the learning process (Hannaford, 1995). Design has an obvious role in helping to create spaces where such richness and diversity of experience can happen – especially for children living in deprived or stressful circumstances.

Taken at face value and as understood in this chapter, the concept of universal design will justifiably include all disenfranchised groups (such as children) whose freedom is currently constrained by environmental barriers, which they are unable to influence or redesign to support their particular needs. It seems obvious that all user needs must be addressed if the design of a space is to be considered 'universal'. Based on the original premise, we may conclude that to be valid, the evaluation of a public environment must address the needs of *all* users, including those with disabilities.

This chapter provides an opportunity to contribute to the discourse through findings from an on-going study of a universally designed park created as an inclusive community environment. Kids Together Park is located in Cary (a fast-growing town with a population of 116,000 in 2006, adjacent to Raleigh, state capital of North Carolina, USA). The park was conceived by the Cary Parks Commission as a family recreation facility accommodating the needs of all users. A community-driven design process was launched in 1994 with a workshop involving children and adults. Children created the name 'Kids Together' and remained an essential part of the design process (see Figure 7.1). After several years of community fund-raising, the park opened in 2000. About one million US dollars were invested, including the cost of extensive infrastructure and site works. The park serves as a research site for systematic studies of park use, including the data reported on here.

7.1 Children's workshop during the community design process for Kids Together Park.

Study goal

Kids Together Park is an appropriate study site because of its design, which offers a diversity of high-quality activity settings potentially attracting multiple user groups. This provides the base condition for an ecologically valid research design.[1] Data for the study reported here were generated using *behaviour mapping, behaviour tracking, park visits with people with disabilities, setting observations*, and *interviews with users*. Using this multi-method approach, the study goal was to learn how a universally designed park was used and perceived. The purpose was to contribute to the extant evidence-based literature on park and playground design (Cooper Marcus, 1990; Moore et al., 1992).

Theoretical framework

Three overlapping concepts provide the theoretical framework for this study:

- territorial range development
- behaviour setting, and
- affordance.

Territorial range development recognizes that maturing children explore, discover and make sense of their expanding world through experience, learned skills and spatial understanding (Hart, 1979; Moore and Young, 1978; Moore, 1989). To maintain this dynamic relationship with the environment, children repeatedly act at their territorial limits, constantly expanding the 'known' world by pressing against the 'unknown'. For each child to exercise her or his exploratory skills beyond the known, space must be designed with soft, extendable territorial boundaries. Given the range of ages, levels of ability, and variety of child–caregiver relationships present in an urban park, environments with higher levels of diversity are likely to satisfy the exploratory needs of more children at any given moment.

Applied to park design, this view of territorial range development provides children with a landscape offering new exploration challenges and discoveries with each visit. A park with effective territorial range development would thus hold a child's interest through repeated visits across the span of childhood. Territorial design must similarly motivate the continuing interest of accompanying caregivers. They must be as excited to go to the park as their children and feel comfortable once they get there.

Behaviour setting is an ecological unit where physical environment and behaviour are indissolubly connected in time and space. Barker (1976) describes behaviour settings as the subspaces of a geographical area and the predictable patterns of behaviour they afford. Behaviour settings are composed of *entities* and *events* (people, objects, behaviour) and dynamic processes such as sound and shade. Their components are arranged functionally as part of the whole. Functions are independent of adjacent eco-behavioural units. The concept is useful for analysing human spaces because it provides a theoretical means to disaggregate their functional parts, thus providing a key structural component and unit of analysis for the interpretation of findings. Empirically established levels of use can be compared to investment and management costs to provide park managers with benefit/cost measures that can be used to shape future management strategies.

Applied to park design, the behavioural setting concept provides an invaluable vehicle for specifying the function of sub-areas and laying them out in appropriate relationships to each other within the whole park. At the level of behaviour setting, requirements to support people with disabilities will be considered with the requirements of all other users.

Affordance is a concept (Gibson, 1979) which defines functional physical features 'that offer certain possibilities to the individual' (Heft, 2001: 297). Affordances are the functional properties of environments related to individual users. They are neither part of the environment, nor of the perceiver. An affordance exists at the intersection of the subject's behaviour in connection with the environment. Potential affordances exist even if the individual has not yet discovered them. It is the individual's action that makes an affordance 'actualized'. Individuals 'pick up' information by perceiving the relation between the layout of the space, objects and events and their developing skills (Gibson, 1979). As children pick up information afforded by the layout, objects and events in behaviour settings and learn the possibilities for action they offer, these actualised affordances become embodied knowledge that support relationships between individuals and environments. Affordance is a dynamic perceptual process through which interrelationships with behaviour settings develop over time. Affordance considers the individual and the environment as an interactive system.

Applied to park design, the concept of affordance can be used to identify and analyse similarities and differences among behaviour settings such as manufactured play equipment, sandplay areas, pathways and vegetated settings. It is also valuable for explaining, in terms of design details, variations in activity across behaviour settings of the same type. For example, the reason why one sandplay setting may be more popular than another for caregivers with young children could be explained by the elevated enclosure for sand that also affords sitting, like a 'sitting wall'. The layout of settings and territories may vary in dimensions such as geometric form, variations in topographic variety or visual transparency. Components may require specific features such as handholds to make them accessible to children. Characteristics of plants such as fragrance or pickable seeds or fruit may influence the actualization of affordances. Natural events, such as weather, or social events, such as birthday parties, may also influence the actualization of affordances.

Empirical evidence identifying affordances can provide valuable source data for designers by focusing attention on the detailed design of components (*layout*, *objects*, *events*, and for designers we may add *features* and *characteristics*) that really matter from the point of view of users. The extent to which such evidence is associated with a particular component of a behaviour setting may disclose a measure of its universal design value.

Application of theory to park design

Together, territorial range development, behaviour setting and affordance should be thought of as closely linked environment–behaviour constructs that provide a theoretical base for measurement of behavioural links between the built environment and physical activity (Gibson, 1979; Gibson and Pick, 2000; Heft, 2001).

If the design of a neighbourhood family park is considered as the task of creating a community meeting ground or commons, support of social, psychological and cultural objectives is of paramount importance. Such a park will serve a longitudinal function as a place where children, families and communities can develop and become sustained for all ages and abilities.

For children, parks can serve as communal backyards, where they can play freely together and be exposed to experiences that may be unavailable in constrained domestic settings. The overall territory of a park secures support for natural child development by allowing safe access to an ever-widening range of experience in both breadth and depth for children alone, with peers or accompanied by caregivers. An appropriately designed park environment will challenge the increasing maturity level of each individual and at the same time respond to parents' differing levels of tolerance towards children's risk-taking. Activity in settings is triggered by the child's increasing repertoire of actualized affordances learned from the potential for action that settings offer. Diversity of settings and richness of child-related features are the design criteria likely to differentiate more successful from less successful territories from the point of view of child development and family usability.

Methodology

A multi-method research strategy was used to assess the park design through a participatory, inclusive approach that regards users' knowledge and behaviour as a valid and appropriate body of data. Three types of data were collected. First, park-wide spontaneous activity data were collected using *behaviour mapping*, *behaviour tracking* and *setting observations* (described below). To expand the theme of social inclusion, informal observations of use of the park by ethnic/racial minorities and adolescents were included. Second, selected families with a member with a disability were recruited to make a *videotaped park visit*. Third, *on-site interviews* were conducted with the above families as well as with other park visitors.

Three levels of analysis were conducted. First, the observational data were analysed to investigate the *pattern of use* in the park as a whole by children and adults, both in terms of its functional zones and types of behaviour settings. This first level of analysis produced an environment-behaviour assessment of park-wide use, including an understanding of how the dynamics of use between settings were influenced by the park layout. In this regard, the function of the composite play structures and primary pathways received special attention.

The second level of analysis introduced further data to help explain the *variations of use* across different types of behaviour settings by children and adults. To contribute an understanding of effective park site use, relationships between size of settings and behaviour were investigated.

The third level of analysis was aimed at understanding *special uses* of the park – how the layout, settings and features of the park landscape afforded satisfying experiences for children with disabilities and other family members. Dominant park perceptual themes of safety, freedom and ambience identified from interviews with park users were also discussed.

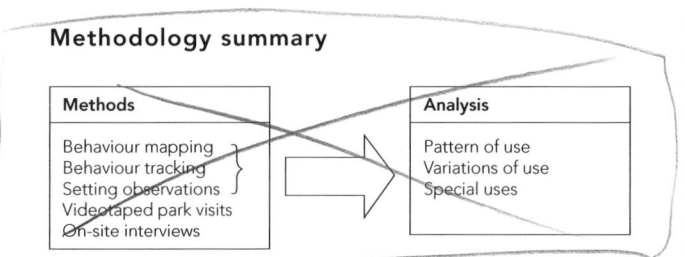

Methodology summary

Methods		Analysis
Behaviour mapping Behaviour tracking Setting observations Videotaped park visits On-site interviews	→	Pattern of use Variations of use Special uses

Procedure

The distinctive layout of Kids Together Park was defined by three intersecting circular pathways that functioned as a behaviour setting type as well as affording access to the other park behaviour settings (see Figure 7.2).

Behaviour mapping, which records the location of use across the site, was conducted by systematically circulating through the space, coding each user by type and location. Behaviour setting boundaries were first established.[2] Coding of users included child in stroller, ambulatory child or adult, wheelchair user (child or adult) and gender (for adults only, because we were interested in caregiver behaviour).[3] Behaviour mapping data were recorded on a paper plan of the site,[4] later entered and processed using Geographical Information Systems software (ArcMap 9.1, ESRI).

Behaviour tracking (a form of behaviour mapping), which records use of the site by single individuals or small groups of individuals, was conducted by following family groups (with their consent) through the space. Each track was recorded on a paper plan (with park entry and leaving times). The routes followed by adults and children were plotted separately.[5] Subjects were treated as a convenience sample. As subjects entered the park they were selected to progressively create a group of trackings covering a range of user types (by family composition, age, ethnicity and gender).

Setting observations were made during the course of multiple walks through the park and conducted by observing the detailed activity of a given setting for the duration of a natural sequence of activity occurring there, usually for several minutes. Observations were noted on a standard form with fields for weather, type/size/age/gender of group(s), type(s) of activity, durations, components of setting used and other observations. Field notes and related photographs were made of user interactions in the setting with physical settings, features, accompanying family members and other park visitors.

Family visits were conducted by first welcoming the family group (families including children with disabilities) at the park entrance to complete consent formalities. After making clear that the group should follow their own path around the park, the behaviour of the target child with a disability was videotaped (including voice captured with a wireless microphone), to record interactions with physical settings, features, accompanying family members and other park visitors. Structured, open-ended interviews were conducted with family members at

7.2 Aerial view of Kids Together Park, soon after construction.

7.3 The functional use zones and behaviour settings of Kids Together Park.

Zones
- Park Entry
- Park Pavilion
- Young Children
- Vertical Composite Structure
- Horizontal Composite Structure
- Dragon Lawn Gathering
- Primary Pathways

Parking settings not included in analysis.

Behavior Setting Types
AT - Anchored toys
CS - Composite structures
GA - Gathering
GR - Grassy
DR - Dragon
PH - Playhouses
PP - Primary paths
SA - Sandplay
SC - Stepbar climber
SP - Secondary paths
SW - Swings
TP - Tertiary path

7.4 Behaviour map of Kids Together Park showing the distribution of child users.

90

WHAT MAKES A PARK INCLUSIVE AND UNIVERSAL?

the end of the visit. Including the interview, visits lasted 60 to 90 minutes.

In addition to the park visit interviews, students in the first author's classes conducted three uncontrolled, *ad hoc* park user interview surveys over a four-year period, totalling 80 individual interviews. Although these data cannot be considered as a systematic survey, they provide evidence of users' dominant perceptions.

Analysis

The GIS relational database was used to estimate spatial distribution of use related to setting type, setting size, child/adult ratio by setting type and gender ratio by setting type. For the purpose of the analysis, behaviour mapping data were distributed between 40 individual behaviour settings, covering a total of 12 behaviour setting types, within seven functional use zones of the park (see Figure 7.3). Figure 7.4 shows the behaviour map for children.

Functional use zones

The functional use zones were defined as follows:

1. *Park Entry Zone* contains five settings: (not included in analysis) two car parking areas and one disabled persons' car parking area, an approach path/accessible route, and (included in analysis) entry plaza/gathering area with benches and tactile map (see Figure 7.5).

2. *Park Pavilion Zone* contains one setting: park pavilion with picnic tables for group gathering (also contains public toilets) (see Figure 7.6).

3. *Young Children Zone* contains ten settings: small swings, two playhouses, sand and water play, secondary path with bridge, little bridge toy, spring toy, lawn patch, seat wall and bench gathering areas (see Figure 7.7).

4. *Vertical Composite Structure Zone* contains eight settings: low hill with vertical composite structure, two lawn patches, pergola gathering area with seats, sitting wall gathering area, wash-off gathering area, secondary path with benches and tertiary path (see Figure 7.8).

7.5 Zone 1: Park entry – approach, parking, accessible route, entry plaza.

7.6 Zone 2: Park pavilion – group gathering.

ROBIN C. MOORE AND NILDA G. COSCO

7.7 Zone 3: Young children – small swings, playhouses, sand and water play, little bridge, grass patch, gathering.

WHAT MAKES A PARK INCLUSIVE AND UNIVERSAL?

7.8 Zone 4: Vertical composite structure – low hill, secondary path, tertiary path, grass patch, gathering.

7.9 Zone 5: Horizontal composite structure – large swings, other equipment settings, sand play, accessible sand play, secondary path.

5. *Horizontal Composite Structure Zone* contains eleven settings: horizontal composite structure, large to-and-fro swings, tyre swing, platform swing, balance beam, stepbar climber, sandplay with digger, raised accessible sandplay, group gathering with picnic tables, group gathering in pine tree grove and secondary path (see Figure 7.9).

6. *Dragon Lawn Gathering Zone* contains four settings: dragon sculpture (Katal – 'Kids are together at last'), surrounding, sloping lawn, and picnic tables and tree grove gathering settings (see Figure 7.10).

7. *Primary Pathways Zone* connects to the Park Entry Zone, accesses each of the other zones and contains four settings, each a path section with benches, sitting walls and drinking fountains (see Figure 7.11).

Zone attractiveness index

To measure attractiveness across zones requires taking into account both the proportional amount of use attracted by each zone as well as the proportional number of settings contained by each. Gross level of use of a zone does not constitute a relative measure of attraction unless moderated by the number of settings within the zone. An index of attractiveness is proposed representing the ratio of the percentage of use

7.10 Zone 6: Dragon lawn gathering – Katal dragon sculpture, sloped lawn, tree grove, picnic tables gathering. Photograph shows chase game afforded by Katal features.

7.11 Zone 7: Primary path – benches, sitting walls, drinking fountains. Photograph shows social interaction afforded by the wide walking surface.

compared to the percentage of settings for each zone, where the value 1.00 is neutral (see Figure 7.12).

Almost two-fifths of the total park use (39.83%) occurred in the Horizontal Composite Structure Zone 5 (see Figures 7.9 and 7.13), with a level of use 2.26 times the next rank order zone (Zone 7, Primary Pathways, see Figure 7.11). Zone 5 was also one of the two most diverse zones as measured by the number of settings per zone, which range from ten (Young Children Zone) to one (Park Pavilion Zone).

Index of Attractiveness values range between 1.64 (Horizontal Composite Structure Zone) and 0.45 (Young Children Zone). Three other zones have ratios above 1.00: Primary Pathways Zone (1.45), Park Pavilion Zone (1.45) and the Dragon Lawn Gathering Zone (1.06). These four zones could be considered the most attractive relative to the number of settings they contain.

The Horizontal Composite Structure Zone (5) was the most attractive with an index score of 1.64. In contrast, the Vertical Composite Structure Zone score was considerably lower (0.54 – less than 1.00). Why was Zone 5 (located furthest away from the park entrance), so attractive?

It is possible to speculate that since this zone (and the Young Children's Zone) contained more behaviour settings (nine and ten respectively) than other zones, it had more potential to attract a broader range of users with different levels of skill and ability. It was also easily approached and accessed because a primary path connected to its two ramped entries. Not only were there more settings in Zone 5, they were also easily accessible. The swings and horizontal composite play structure components (ramps, slide, overhead glider) were accessible directly from the adjacent primary pathway, which served as a circulation and access spine for a variety of play options as users moved around the space (see Figure 7.14a). These enabled caregivers with strollers to penetrate the setting to use the shady gazebo with comfortable seats, which afforded social gathering within the structure. The upper platform offered a vantage point for caregivers to supervise their children within

ROBIN C. MOORE AND NILDA G. COSCO

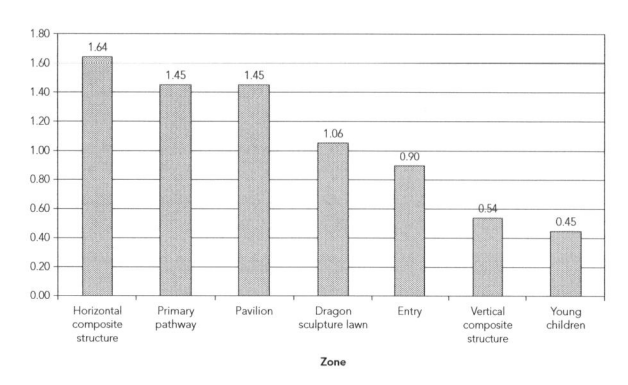

7.12 Attractiveness index of zones (percentage of use/percentage of settings).

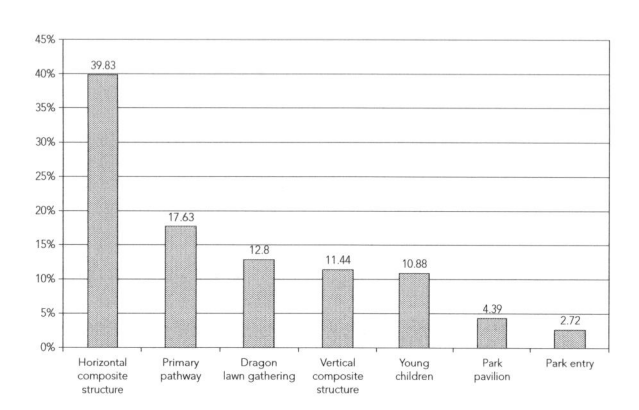

7.13 Percentage of use by zone.

the zone. Caregivers with children in strollers could relax, observe what was going on around them from an elevated position, and participate visually and aurally in the activities of other family members, including older siblings. It can be anticipated that wheelchair users could also benefit from the elevated gazebo setting; however, none were observed in this zone.

Setting observations showed extended family members (grandparents, aunts, uncles, etc.) interacting more with children in the other settings of Zone 5 than in other park zones (see Figure 7.14b).

7.14a Diversity of user interaction afforded by the proximity of the primary path to the variety of features of the horizontal composite structure.

7.14b Diversity of user interactions afforded by swinging settings on the opposite side of the primary pathway from the horizontal composite structure.

WHAT MAKES A PARK INCLUSIVE AND UNIVERSAL?

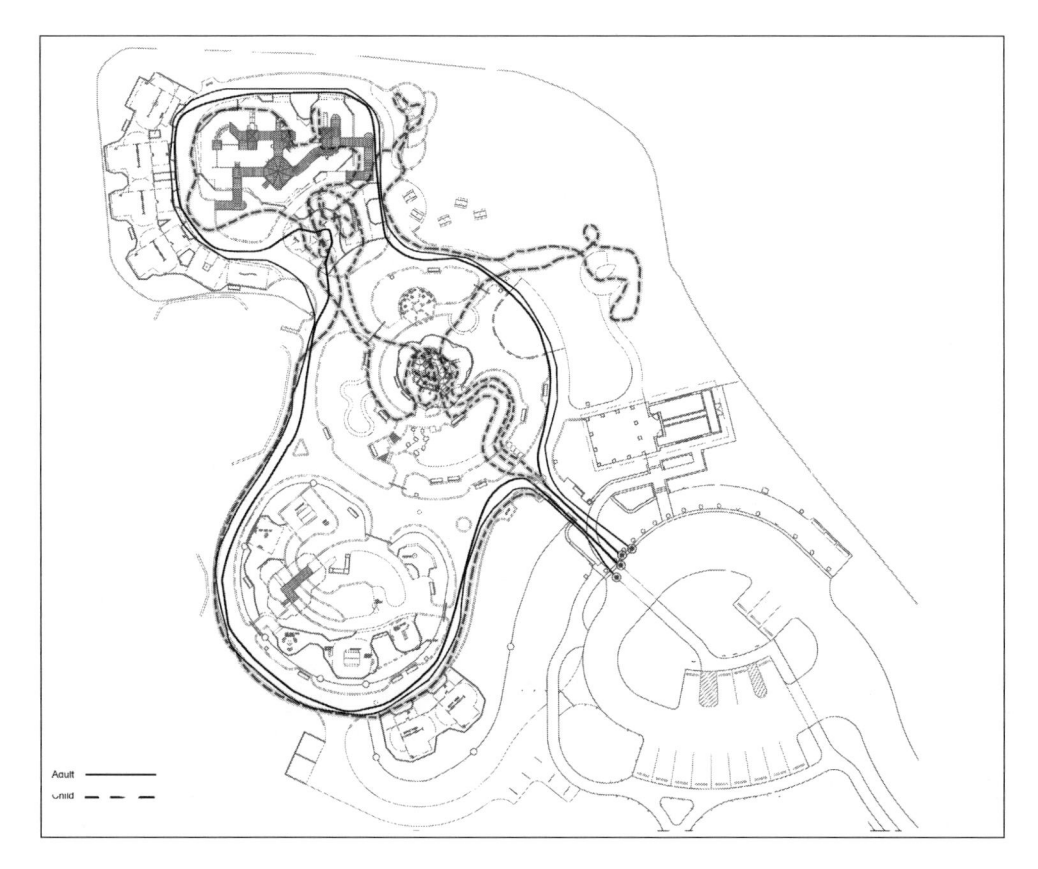

7.15 Tracking record of parents and child showing the parents strolling around the looping paths, while their child played in adjacent settings, periodically rejoining parents.

One of the individual trackings demonstrated the particular behaviour pattern of children darting off to play in adjacent settings, while accompanying adults moved along the path (see Figure 7.15). This pattern was most pronounced around the Horizontal Composite Structure Zone because of the larger number of adjacent play opportunities. The sense of seamless connection between primary pathway and adjacent settings in Zone 5 was visually reinforced by the distribution of vegetation, which penetrated both the horizontal composite structure and the swing settings on either side of the primary pathway.

Park Pavilion Zone. Even though the Park Pavilion Zone (see Figure 7.6) was a single setting mainly accommodating family gatherings such as birthday parties, it attracted 4.39% of use, which explains its relatively high attractiveness index.

The Primary Pathway Zone. Subdivision into four settings (12.12% of the total number), each serving adjacent zones and accounting for 17.63% of use, also gave the Primary Pathway Zone a score of 1.45. Why was this zone so attractive?

Many of the on-site interview respondents mentioned the generous width (3m/10ft) of the pathways that afforded easy movement through the park, especially for larger family groups with children riding wheeled toys (see Figure 7.11). This subtle dimension of inclusion provides young children with space to energetically move with less risk of conflict with other users or causing anxiety to caregivers. Respondents also noted the curving form of the pathways that progressively exposed the landscape, adding visual interest to the pedestrian experience.

Kids Together Park demonstrated how pathways can be designed to provide a movement armature throughout the park for strolling and informal socializing in the tradition of the *paseos* in Spain or *promenades* of France and England. Wide, curving paths afford inclusion because a group of half-a-dozen or so can walk and chat together, allowing lulls in conversation to be filled by attention to the progressively exposed sensory landscape and activities of other users – that may stimulate further topics of conversation. Inclusive social relationships are constrained by narrow paths where individuals must walk behind each other or break the conversation to make way for groups coming in the opposite direction. For groups containing children in prams or strollers or wheelchair users, wide pathways are especially beneficial.

Dragon Sculpture Lawn Gathering (Zone 6). With a score of 1.06, this zone was ranked fourth in attractiveness. It contained four settings (12% of total number, Katal, lawn, tree grove gathering and picnic table gathering) and accounted for 13% of total site use. Activity was mostly related to Katal. The evocative creature attracted children to the zone, who could climb and chase around the dragon and the adjacent sloped lawn, engaged in gross motor activities such as rolling, and activities with caregivers such as wheeled toy and ball play (see Figure 7.16). Other caregivers were able to gather around the adjacent picnic tables settings with their children close by, playing on Katal and sloped lawn. This relationship was observed especially when the picnic tables were used as a base for a birthday party or other family gathering event.

Distribution of use by setting type

Distribution of the behaviour mapping data across the 12 setting types allows a more highly differentiated level of analysis of use than for functional zones. Park settings and components

7.16 Intergenerational play afforded by the sloping surfaces around Katal.

WHAT MAKES A PARK INCLUSIVE AND UNIVERSAL?

afforded movement (walking, running, climbing, rolling, hiding-and-seeking, sliding, swinging) on and around manufactured equipment, pathways, topography, trees, shrubs and ground surfaces, and socialising (talking, partying, being with others, observing others) on custom-designed benches, sitting walls, picnic tables and in a pergola and park pavilion. Distribution of this pattern of use by setting type provides an overall environment-behaviour measure, which indicates relative park use across the site from the most to least used setting types. Distribution of use by setting type can inform discussion about the social implications of park design. Equally, empirical findings can better inform physical design to support desired social outcomes.

Of the twelve park behaviour setting types coded, four (composite play structures (25.67%), swings (14.87%), primary pathways (13.82%) and gathering areas (12.20%)) accounted for almost two-thirds (66.56%) of the use. The addition of sandplay (10.10%) indicates more than three-quarters (76.66%) of use occurring in five setting types (see Figure 7.17).

Overall, these findings suggest that park users were attracted by the areas with manufactured play structures, including swings and sandplay, the varied gathering settings (benches designed as art objects, park-style benches, sitting walls and group sitting areas) and the primary pathways. The relatively high use of gathering and pathway settings indicates the social attraction of the park.

Informal observations of gathering settings indicated a variety of user group configurations including groups of parents chatting in the Young Children Zone, couples using benches, family picnics in the picnic tables and park pavilion settings, and single individuals reading on the sitting walls and benches. While chatting adults strolled through the park, their children played in adjacent settings or engaged in chase games with each other and with adults on the primary pathways wide enough to accommodate active play without disturbing other users (see Figure 7.15).

Setting type user profiles

So far, the analysis has focused on use patterns at two levels of environmental subdivision (zones and behaviour settings), without differentiating user types. Behaviour mapping included type of user (child/adult), adult gender, and the presence of strollers and wheelchairs. These data provide two additional use distribution measures by user subgroup.

Child/adult ratio (CAR) is an index of the extent to which different types of behaviour setting are used by adults, children or mixed groups. In other words, where do children and adults play together or separately? CAR is calculated by dividing the proportion of child users by the proportion of adult users for each setting type. A value of 1.00 indicates equal use. A value greater than 1.00 indicates child dominance. A value less than 1.00 indicates adult dominance.

Figure 7.18 shows the proportion of use between children and adults across setting types. Katal the dragon was the most strongly child-attracting (CAR 3.15) by a factor greater than 3:1, followed by sandplay (CAR 2.35). The combined CAR for the two composite structures was 1.74 (mostly due to the vertical structure, with a CAR of 3.32 compared to a CAR of 1.41 for the horizontal structure), thus supporting the earlier discussion about the ease of access of the horizontal structure by adults compared to the vertical structure. The stepbar climber CAR of 1.63 indicates the difficulty of access onto the structure for adults. Their presence was observed helping and supervising their children on the equipment.

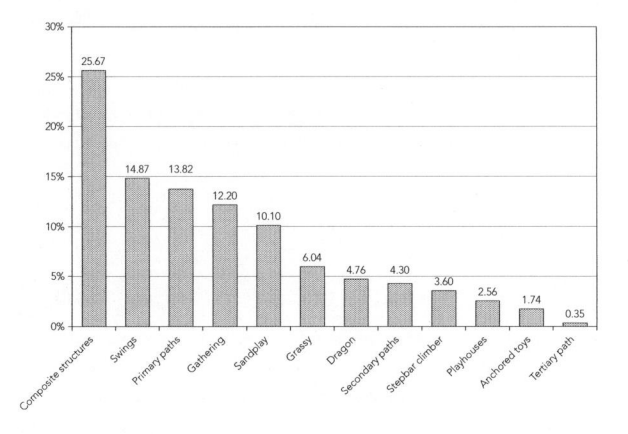

7.17 Percentage of children's use by behaviour setting.

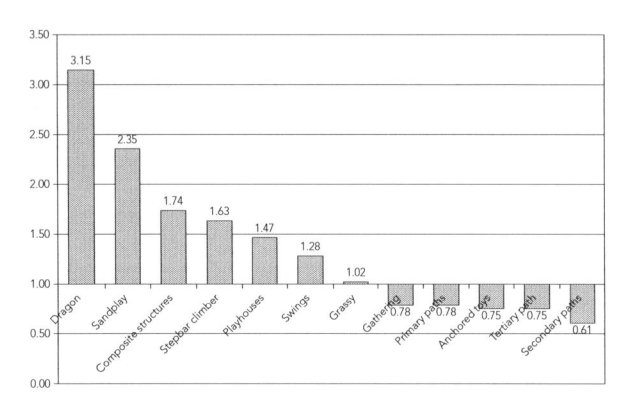

7.18 Child/Adult Ratio (CAR) by behaviour setting.

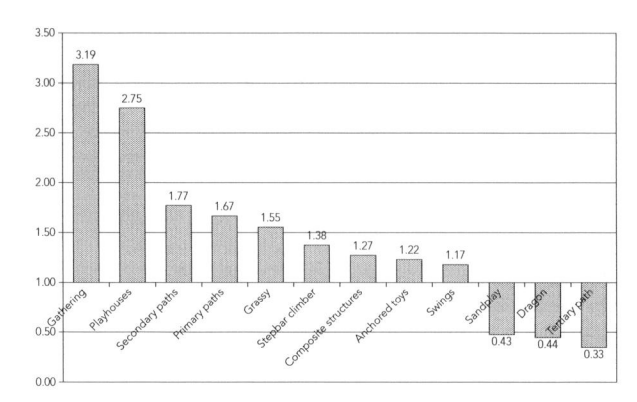

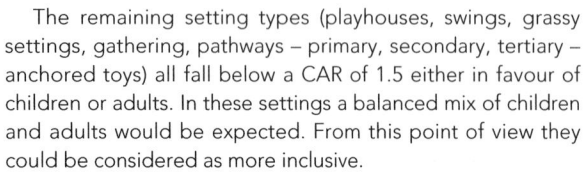

7.19 Female/Male Ratio (FMR) by behaviour setting.

The remaining setting types (playhouses, swings, grassy settings, gathering, pathways – primary, secondary, tertiary – anchored toys) all fall below a CAR of 1.5 either in favour of children or adults. In these settings a balanced mix of children and adults would be expected. From this point of view they could be considered as more inclusive.

Female/male ratio (FMR) is a measure of the extent to which different types of behaviour setting are used by adult females, adult males or mixed adult groups. In other words, where do women and men gather together or separately? FMR is calculated by dividing the proportion of adult female users by the proportion of adult male users for each setting type. A value greater than 1.00 indicates female dominance. A value less than 1.00 indicates male dominance.

Figure 7.19 shows the proportion of use between women and men adults across setting types. Gathering areas are clearly the most dominant female settings with an FMR greater than 3. It is interesting to note that similar gender-differentiated 'social gathering' behaviour was identified in a behaviour mapping study of 5-to-9-year-old children conducted in a diversified schoolground (Moore and Wong, 1997). The playhouses are a close second in adult female dominance with an FMR of 2.75. This female dominance may be explained by the setting observations of female caregivers engaged in dramatic play with domestic themes with their children.

Spatial distribution of use

The findings thus far have focused on the distribution of use in terms of aggregate users and user groups across types of settings. But this leaves out the crucial variable of space as measured in square metres/square feet.

Mason *et al.* (1975) used behaviour mapping (and user interview data) to justify the importance of small neighbourhood parks in Berkeley, California. We know of no other study used to measure site and setting effectiveness across a whole park system using behaviour mapping. Data from this unpublished study were analysed by Moore (1989) to develop measures, some of which are used in the study reported here. Unfortunately, use data on other parks in the Cary system were not available as part of the present study so inter-park comparisons cannot be made. However, the KTP behaviour mapping data enable an intra-park comparison to be made across settings using the use/space ratio measure developed by Moore and Wong (1997) and Moore (1989).

Use/space ratio (USR) measures the amount of use in relation to the size of behaviour settings (percentage of total use of each behaviour setting divided by the percentage of total area of all settings). Figure 7.20 shows that, from this point of view, composite structure settings and sandplay settings, with USR values of 2.19 and 2.17 respectively, are the most effective setting

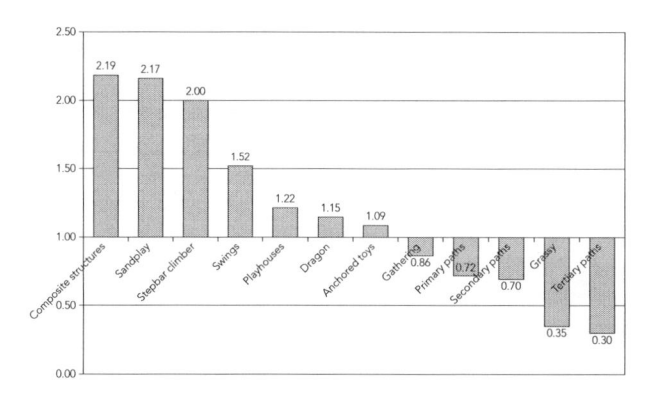

7.20 Use/Space Ratio (USR) by behaviour setting.

types measured by the amount of activity attracted in comparison to their size. Scores for the stepbar climber (USR 2.00) and swings (USR 1.52) also indicate effective space use. Values for Katal (USR 1.15), Playhouses (USR 1.22) and Anchored Toys (USR 1.09) are also on the positive side. Although the composite structures had a combined score of 2.19, the individual scores were markedly different. The vertical structure had a relatively high USR of 3.10 because the setting footprint was small compared to the amount of activity attracted. In contrast, the spread out horizontal structure had a USR of 1.97 because of the larger footprint.

Use by children with disabilities

Park visits were arranged with families with a child with a disability. A group of children with sight disabilities was also observed visiting the park. Four visits are reported here that provided an opportunity to observe the uniqueness of individuals with different impairments, in the context of family, responding to the opportunities of a diverse physical environment offering a broad range of behavioural choices. The visit summaries presented below illustrate affordances that appear to be primarily 'sensory'. To say so expands the use of the term 'affordance' into a broader current discourse concerning different possible types of affordance (Hartson, 2003). Since affordance was originally formulated as a concept of perceptual psychology, to consider it from the sensory point of view of body-in-space seems a justifiable step, including the three inter-related component senses: the *kinaesthetic* (sense of *movement* through space); the *vestibular* (sense of *balance* in relation to the force of gravity); and the *proprioceptive* (sense of the *position* of body and limbs in space). The following descriptions illustrate how individuals with a variety of disabilities can discover body-in-space sensory stimulation afforded by a diverse range of settings.

Visit 1 – the challenge of horizontal movement. This informative visit demonstrated how a variety of undulating, curving pathways and shallow steps afforded challenges to someone with low muscle tone. The 28-year-old, almost nonverbal daughter arrived in a wheelchair; however, the first thing the mother did was to make her get out and ambulate. 'Let's get out of the wheelchair and walk; it is good for your mobility,' she said. Together with a family friend, they played on one of the Talking Benches (interactive art objects in the entry plaza, made of curly, steel talking tubes with mouth/ear pieces at each end), which afforded a fun moment of rudimentary verbal interaction.

Afforded by the wide path, the mother pushed the wheelchair ahead so the daughter (reluctantly) had to run after and catch it. The verbal interaction continued, with the mother intent on encouraging her daughter to exercise as much as possible. The daughter pushed the wheelchair like a 'walker' on what became a psychomotor challenge course through the Young Children Zone (see Figure 7.21), pushing the chair up the curving ramp, across the bridge, navigating a sharp bend and chasing the chair down the other side.

As they entered the Vertical Composite Structure Zone, the mother dragged the wheelchair up the wide, shallow stone steps while patiently coaxing her daughter up, one step at a time. The daughter's low muscle tone meant that the 10–13 cm (4–5 in) risers were challenging. They took several minutes to climb with the mother's loud words of encouragement.

The daughter was clearly apprehensive about using the vertical composite structure. She did not seem to understand how to use the transfer platform. The mother, friend and one of the researchers together helped the daughter navigate the steps up to the first level tunnel, through which the daughter

ROBIN C. MOORE AND NILDA G. COSCO

7.21 Pushing a wheelchair up the secondary pathway ramp to the bridge over the 'river of sand' in the Young Children Zone.

was pulled feet first. She appeared insecure, even though the transparent tunnel was not much longer than she was. This contrived, overly challenging experience was not enjoyed by anyone. The interior space of the structure afforded an easier route to navigate. With assistance, the daughter mounted the interior platform (50 cm/20 in) above the woodchip ground surface) and was pulled through the short connecting tunnel to the outside. She laughed and seemed to enjoy the experience. By now it was obvious that the vertical structure did not match the daughter's abilities.

The horizontal composite structure was a different story. As the daughter was tired, she got in the wheelchair and was pushed by her mother up the long entry ramp. They stopped at the slide at the higher level but the entry platform was too high to climb to get to the slide itself. They tried the lower slide but the daughter was very apprehensive and the plan was abandoned. Instead, the mother raced the wheelchair and daughter up and down the ramps and through the structure, simultaneously making loud motor noises. The daughter smiled and laughed, expressing enjoyment. Back at ground level, the daughter was able to climb on the webbing net suspended below an upper platform and, after considerable encouragement, was brave enough to allow herself to 'fall down' on the soft, bouncy surface. A repeat performance was too challenging.

Visit 2 – sibling can facilitate swinging enjoyment. This visit demonstrated the role of an able-bodied sibling in facilitating the enjoyment of vestibular stimulation[6] afforded by a variety of swinging devices (Ayres, 1998). The family group included mother, father and two daughters – one able-bodied, the other her 14-year-old, developmentally disabled, nonverbal younger sister – and this sister's caregiver. The younger girl was attracted to the tyre swing and enjoyed watching children using it (see Figure 7.22). The older girl commented that the tyre swing allowed her sister to get close to the other children, to feel part of the action. The older sister got into the tyre swing by herself

WHAT MAKES A PARK INCLUSIVE AND UNIVERSAL?

7.22 Swing activity affords social interaction between children with and without disabilities, including family members.

so that her sister could push her. Other children joined her in the tyre, while her sister continued to communicate through her body language that she felt part of the action.

The family moved to the cradle swing (a wide, moulded plastic form provided for children who do not have the ability to sit up and grip the swing chains). The cradle swing is popular with all children because it provides a different experience (prone, looking up at the sky) to a conventional to-and-fro swing. The height of the cradle made accessibility challenging for the younger girl. With her sister's help, she eventually slipped into the seat and appeared to enjoy the rocking sensation (vestibular stimulation).

At the platform swing (square, spring-mounted, metal platform with a central post enabling users to rock back and forth or follow a circular rocking motion), the sisters mounted the platform and both held on to the central post, which the older sister operated, so they could play together. Again, the motion evidently produced enjoyment.

Visit 3 – the pleasure of swinging in secluded natural surroundings. This visit demonstrated again the important role of a close relation (in this case the father) in facilitating swinging. Father, mother and son (43-year-old, autistic, nonverbal, ambulatory) headed straight to the to-and-fro swings and spent the bulk of the time there. The son clearly enjoyed the vestibular stimulation of swinging and was able to pump himself. The father used the adjacent swing to accompany his son (see Figure 7.23) and said they spent a lot of time outdoors together, especially in natural areas, which his son enjoys. He commented that 'my son gets anxious when too many people are around so it is good to be in a place where escape to a more secluded setting is an option' (the to-and-fro swings feel secluded because they are located against the park boundary fence and are separated from the main path by a line of shade trees).

Visit 4 – 'It's like a big playroom.' This visit involved a visiting group of four children of 8 to 10 years old, all legally classified as blind. Accompanied by their caregivers, the four children

7.23 Father and son swing together, while mother looks on.

moved excitedly through the park settings with surprising ease and obvious enjoyment. They especially enjoyed settings that afforded vestibular stimulation (swings, slides, overhead glider), kinaesthetic stimulation (corkscrew slide, fireman's pole) and proprioceptive stimulation (tunnel/bridge, ramped route through the low structure).

Lacking sight, the children's proprioceptive sense especially appeared more developed or at least more central to enjoyment of body awareness as they moved in, on or through varied three-dimensional spaces. Observations of this group (including children blind from birth) reinforced the notion of enjoyment that can arise from being able to 'read' the three-dimensional qualities of space in terms of its bodily affordances – 'Like a big playroom', as one child said.

Park visit commentary

At the end of the visit, the family visitors were asked what they found most attractive, for suggestions for improvement, and to comment on the park as a whole. What visitors liked most included: 'The low structure with ramps and ups and downs is easy with a wheelchair.' 'The park works for wheelchair users.' 'The path structure has a nice flow, easy to wander around.' 'Swings! Tyre swing.' 'Flowers to smell. Plantings. Foliage is beautiful.' 'Benches to watch people.' 'The dragon and "pool"' (water gathered in the dragon's 'tail' after a rain). These comments suggest that the three-dimensional flow, choice of swinging opportunities, and flowering shrubs are the most attractive attributes of the park for families with a child with a disability.

General family comments about the park included: 'Attracts people of all ages and abilities.' 'Good for playing.' 'Compact. Feeling of closeness.' 'Intricate complexity is attractive.'

'Aesthetically appealing. Park is different every time you are here.' 'Whole family can play together. Feels like a "family park".' 'Very active. Easy to wander around and get exercise. Nice flow, no dead-ends.' 'You need to go! Beautiful.' 'Unique. Nicer than the bare openness of many playgrounds.' 'Attracts all people, kids and adults. Addresses needs of children. Engages kids.' 'Great park for kids and families with kids.' 'Intergenerational – whole family, old and young, little children. Great place for teens.' 'Nice place for picnicking.' 'Layout is great. Diverse, holds attention. Fenced in. Safe and challenging.' 'Learning experience. It's a natural experience. It is a wonderful park.'

Suggested improvements included additional handrails and handholds in the play equipment, additional swings to reduce waiting, installation of more benches, provision of summer shade, addition of acoustic instruments and fragrant settings and the addition of a family bathroom. Drinking fountains and a water play fountain were highlighted – children were not strong enough to operate them. Larger scale water play settings were desired. An 'ice cream stand' was requested (since implemented). A Braille map and signs to identify the dragon (Katal) were suggested. A blind child asked for 'baby dragons to play with' so he could understand what the big dragon was like (an idea that all children would appreciate).

In summary, comments suggest that an easily navigated, three-dimensional flowing territory, offering a compact diversity of accessible activity choices – including social settings and swinging opportunities – for extended families in an aesthetically appealing, natural environment are the attributes of a park that families with a child with a disability would find most attractive.

Getting to the park

Park visit interviews (including those conducted with *adults* with disabilities not reported above), indicated that car access to the park was straightforward, with 'handicapped' parking situated a few yards from the drop-off/entry plaza zone. From there, broad, almost flat, gently curving, hard-surfaced pathways provide an accessible route to all zones and main settings of the park. However, from the point of view of information and transportation, substantial issues were identified.

Information. Visitors commented that it was difficult to find full information about the park on the Town of Cary website (no one interviewed had discovered the park that way), which gives no sense of the park's uncommon design. Visitors said they discovered the park by word of mouth or in the news. The lack of public information about the universal design character of the park may explain the fact that, out of the total of 1,616 behaviour mapping data points, only two observations of wheelchair users were made.

Transportation. The park is not served by public transportation and is therefore not accessible to families without a car or to adults who can't drive because of a disability or lack of resources.

Cultural inclusion

The picnic tables, pine grove and park pavilion were used for family gatherings, including birthday parties. These behaviour settings allowed flexibility for *ad hoc* user-defined ethnic traditions. One afternoon, a mixed-age, extended Asian family of ten or so were observed in the pine grove, picnicking on blankets spread on the pine needles rather than sitting at the picnic tables (see Figure 7.24). Some days later, a Caucasian family set up a small shade structure with 'Happy Birthday' banner and organized a birthday party on the picnic tables. Another afternoon, a Mexican birthday *fiesta* was held in the park pavilion, complete with loud musical accompaniment from a CD player, *piñata*, and portable barbeque (surreptitiously tucked around the back of the building). Other visitors could be seen smiling and moving to the beat of the music, indicating enjoyment of the overtly expressive immigrant culture and acceptance by the more sedate established culture. Such activities were an indicator of park family friendliness and an inclusive environment, where groups with differing cultural traditions felt comfortable and accepted by the majority culture.

To conduct a rough test of this hypothesis, the list of reservations for the Park Pavilion for the 2006 calendar year was obtained from the Cary Parks and Recreation Department and coded for non-English family names. They represented 29% of the total. In comparison, the 2006 'non-Caucasian' population was estimated to be 18% by the City of Cary. The difference

7.24 Asian group picnicking on blankets in the pine grove.

between these values suggests that ethnic groups in the community find the park to be more attractive than would be predicted by the proportion of minority ethnic groups in the population.

A 'cool' adolescent destination

Several setting observations of adolescents supported the comment earlier about the park being a 'great place for teens'. Apart from occasional adolescent couples wandering around holding hands, groups of two to four girls were observed 'hanging out' in the park, sitting talking, swinging, walking around. Adults mentioned that adolescents regarded the park as 'cool'. Given the lack of legitimised settings for adolescents in the public, urban realm, the park may serve as a legitimate, safe, social setting for these much-maligned groups, where they can blend in unnoticed. Further research could investigate specific settings, components and characteristics that may explain why Kids Together Park is attractive to adolescents.

Perceptions of safety

In addition to the park visit interviews reported above, *ad hoc* park user interviews provide a sense of dominant perceptions. 'Safe' was the most frequently mentioned attribute. By safe, users typically meant socially secure rather than physically safe play equipment (a dominant theme in park design and management for many years). 'Wonderfully safe because everything is enclosed,' a mother said. 'You don't have to worry about where your children are.' Other parents reinforced the perception of safety with comments such as 'easy to follow kids around', 'easy to see where kids are', 'location of equipment allows easy supervision'. One mother remarked that the park pavilion was the best position for overlooking the entrance so she could make sure her child didn't wander out of the park. A single, visible entrance is one of the primary principles of defensible design (Newman, 1972, 1975). Physical safety was rarely mentioned. Comments such as 'user friendly', clean and 'beautiful, like out of a magazine' could be interpreted to mean that physical safety was assumed to be covered in an environment perceived as high quality.

7.25 Family members relax in the Young Children Zone.

Freedom and control

Because most parents praised the park as safe, one can speculate that the positive atmosphere and diversity of play opportunities of the park served as a model to help parents to allow active, free play without close supervision. If parents feel secure they will be more inclined to encourage their children to explore, to push themselves as they engage with the environment. Too much or constant parental supervision in the name of safety and security can sometimes result in a loss of play opportunities for children. A child who is continuously told to 'be careful' or directed how to navigate or interact with particular settings will lose the advantage of self-learning, skill building, competence and growth in confidence that results from free play under the child's own volition (Frost et al., 2001). In KTP, over-protective parents were rarely observed.

Ambience – an elusive quality

In interviews, users mentioned being attracted by the overall ambience of the park, especially related to its naturalistic character and richness of planting around the play settings. The positive social atmosphere was also recognized. Users noted as positive the diversity of other users by age, ability, cultural background and gender. Visitors' comments indicated that they did not use the park to escape from other people but rather to enjoy the feeling of community. This was especially evident in the Young Children's Zone. Groups of parents were often observed gathered on the elevated bridge chatting, keeping an eye on their children (see Figure 7.25).

User comments suggest that they enjoyed the inclusive feeling of the park because it had attributes of both social and physical ambience. Visitor attention could be directed to one or the other or both simultaneously. When few visitors were around, the natural ambience was there as an antidote to boredom.

Conclusion – a new public role for inclusive, universal design

High quality, family play area environments are crucial vehicles for inclusion because children's play is such a powerful means of communication – both between children and between children and adults. High quality family play environments can stimulate free flowing, positive interaction among park users of all kinds. The KTP study findings indicate a park that attracts multi-age, multicultural, multi-ethnic/racial user groups who find there satisfying experiences. The research techniques applied in this study can be used to understand objectively how park environments and settings are used and by whom. Together they can serve as a tool to better design and manage scarce parkland resources. Over time, the information generated can be used to affect long-term policy changes to improve park environment quality to better serve users.

The concept of universal design, which includes lifecycle issues such as declining abilities with age, is considered by some experts to embrace a broader social inclusion focus on user groups unable to express their environmental needs because of being excluded from the processes that govern the planning, design and management of the built environment (Drache, 2001). There was a time when adults with disabilities were such a group, who struggled for years to become enfranchised, finally to succeed through the passage of the American with Disabilities Act (ADA) in the USA. Although their struggle to participate fully in civil society is not over, at least the law is an unequivocal ally. Other user groups with particular environmental needs do not have this *legal* advantage and remain largely ignored. Pedestrians and bicycle riders are examples (at least in the USA).

In the case of children, their situation is weaker because they depend on the decisions of adults. The assumption is that if the environment is universally designed, adults will be more inclined to use them and, therefore, children will benefit from the accommodation and their inclusion will be guaranteed.

Social inclusion can be applied as a concept to any group whose needs are excluded from decision processes related to the planning, design and management of the built environment. The concept can move our thinking beyond 'integration' (people of different abilities occupying the same space) to a point where the users of a space feel they are participating in a shared social and psychological world. Inclusive behaviours are those that link people of all abilities, ages, ethnic/racial groups and cultures in positive relationships. Until now, universal design has focused its creative energies mostly on the design of buildings and products. The objective, systematic research techniques used here indicate a new potential for the field to broaden its scope, to move beyond the context of private spaces and consumer products into the public realm of urban places. New, smart data-gathering tools now make it easier to code behaviours, user characteristics and environmental interactions. Richer, more substantial data sets, analysed quantitatively, promise to improve understanding of environment-behaviour dynamics. Designers and managers of urban parks will have new types of objective evidence to help improve the fit between the built environment and users' needs across the community. The design and re-design of urban community parks may represent a major opportunity for implementing this ideal in the years to come.

Acknowledgements

The project that provided much of the data for this chapter was supported by a grant from the United States National Institute for Disability and Rehabilitation Research (NIDRR), administered through the Centre for Universal Design (CUD), NC State University. Laurie Ringaert, former CUD director, was instrumental in supporting the project. Without her enthusiasm and commitment to developing a universal design evidence base for outdoor urban environments, this project would not have happened. Thanks to (then) PhD research assistants Sheila Akinleye, Daryl Carrington, Evrim Demir, Marcelo Guimaraes and Mine Hashas (all now graduated) and current PhD research assistant M. Zakiul Islam, who re-worked the data. We gratefully acknowledge the families who donated time to come and play in the park. Without their interest and willingness to participate and to contribute their insights, this study would have lacked much of its richness. Of course, if Kids Together Park did not exist, this story would never have been written. Literally hundreds of people were involved in creating KTP, too many to mention by name except Cary Parks and Recreation

Commissioner, Bruce Brown, who developed the original concept, the Cary Parks and Recreation Department staff who followed through with the idea, landscape architects Little and Little who executed the design, and citizen Marla Dorrell, who not only midwifed the birth of KTP but also still leads its continuing evolution.

Notes

1. By ecological validity, we mean that the overall designed environment contains a sufficiently diverse range of settings that a study of the variability of human response would be worthwhile and produce useful results.
2. Boundaries of behaviour settings were established prior to the study reported here, based on the results of two pilot studies conducted during the principal author's graduate course: *Human Use of the Urban Landscape*. It might also be noted that, as the park was a tightly defined designed landscape, the large majority of behaviour setting boundaries were defined *de facto* by physical lines in the park layout. This would not be possible in a more loosely designed or natural space, where an initial wave of several cycles of behaviour maps would be required to establish setting boundaries.
3. Behaviour mapping observations were conducted by pairs of observers following predetermined circuits through the space, with one observer travelling clockwise and the other anticlockwise. A single circuit of observation was defined as a round or single layer of activity. All rounds of observation on a given day were defined as a cycle of observation. Multiple cycles were completed covering all days of the week and weekends until all behaviour settings were covered. The total number of cycles were collapsed to produce the complete behaviour map.
4. Since gathering the data for this study, the Natural Learning Initiative (NLI) at North Carolina State University has developed a more efficient and practical method of gathering data using a Personal Digital Assistant (PDA) with pull-down menus. The only item still coded on the paper plan is user location.
5. Since gathering the data for this study, NLI has developed a more powerful method of tracking behaviour by coding video records using The Observer software (Noldus, 2002). This method enables coding of any number of behavioural attributes in parallel time-stamped tracks. The authors acknowledge the work of Daryl Carrington, PhD, who carried out the series of behaviour trackings included here.
6. The vestibular sense is located in the inner ear. One type of receptor responds to gravity when the head is moved.

References

Ayres, J. (1998) *Sensory Integration and the Child*. Los Angeles: Western Psychological Services.

Barker, R. (1976) 'On the Nature of the Environment', in H. Proshansky, W. Ittelson and L. Rivlin (eds) *Environmental Psychology: People and their Physical Settings*. New York: Holt, Rinehart & Winston, pp. 12–26.

Chisholm, S. (2001) *Housing and Social Inclusion: Asking the Right Questions* [Electronic Version]. Canadian Council on Social Development. Available at www.ccsd.ca/subsites/inclusion/bp/sc.htm (retrieved 28 November 2006).

Cooper Marcus, C. (1990) 'Neighborhood Parks', in C. Cooper Marcus and C. Francis (eds) *People Places: Design Guidelines for Urban Open Space*. New York: Van Nostrand Reinhold, pp. 69–115.

Disability Discrimination Act 1995 (c.50) (1995). Queens' Printer of Acts of Parliament, United Kingdom.

Donnely, P. and Coakley, J. (2002) *The Role of Recreation in Promoting Social Inclusion*. Toronto: The Laidlaw Foundation.

Drache, D. (2001) *Social Inclusion? So Much Effort So Little Effect: Do We Need to Rethink the Public Domain?* [Electronic Version]. Canadian Council on Social Development. Available at www.ccsd.ca/subsites/inclusion/bp/dd.htm (retrieved 28 November 2006).

ESRI. Geographic Information System GIS. Available at www.esri.com (retrieved 25 February 2006).

Freiler, C. (2001) *From Experiences of Exclusion to a Vision of Inclusion: What Needs to Change?* [Electronic Version]. Canadian Council on Social Development. Available at www.ccsd.ca/subsites/inclusion/bp/cf2.htm (retrieved 28 November 2006).

Frost, J., Wortham, S. and Reifel, S. (2001) *Play and Child Development*. Upper Saddle River, NJ: Merrill Prentice Hall.

Gibson, J. (1979) *The Ecological Approach to Visual Perception*. Boston, MA: Houghton-Mifflin.

Gibson, E. and Pick, A. (2000) *An Ecological Approach to Perceptual Learning and Development*. New York: Oxford University Press.

Hannaford, C. (1995) *Smart Moves: Why Learning is Not All in Your Head.* Arlington, VA: Great Ocean Publishers.

Hart, R. (1979) *Children's Experience of Place.* New York: Irvington Publishers.

Hartson, R. (2003) 'Cognitive, Physical, Sensory, and Functional Affordances in Interaction Design', in *Behaviour and Information Technology*, vol. 22 (5), pp. 315–338.

Heft, H. (2001) *Ecological Psychology in Context: James Gibson, Roger Barker, and the Legacy of William James's Radical Empiricism.* Mahwah, NJ: L. Erlbaum.

Mason, G., Forrester, A. and Hermann, R. (1975). *Berkeley Park Use Study.* Berkeley: University of California.

Moore, R. (1989) 'Playgrounds at the Crossroads: Policy and Action Research Needed to Ensure a Viable Future for Public Playgrounds in the United States', in I. Altman and E. Zube (eds) *Public Places and Spaces.* New York: Plenum, pp. 83–120.

Moore, R. and Young, D. (1978) 'Childhood Outdoors: Toward a Social Ecology of the Landscape', in I. Altman and J. Wohlwill (eds) *Human Behavior and Environment* (vol. 3: Children and the Environment, pp. 83–130). New York: Plenum Press.

Moore, R., Goltsman, S. and Iacofano, D. (eds) (1992; 2nd edn) *The Play For All Guidelines: Planning, Design and Management of Outdoor Settings for All Children.* Berkeley, CA: MIG Communications.

Moore, R. and Wong, H. (1997) *Natural Learning: The Life History of an Environmental Schoolyard.* Berkeley, CA: MIG Communications.

NEA National Endowment for the Arts (2003) *Envisioning Universal Design: Creating an Inclusive Society* [Electronic Version]. Available at www.nea.gov/resources/Accessibility/ud/issues_b.html (retrieved August 2006).

Noldus. (2002) *The Observer.* Noldus Information Technology: Wageningen, The Netherlands.

Newman, O. (1972) *Defensible Space.* New York: Macmillan.

Newman, O. (1975). *Design Guidelines for Creating Defensible Space.* Washington, DC: US Department of Justice, Law Enforcement Assistance Administration.

Ostroff, E. (2001) 'Universal Design: The New Paradigm', in W. Preiser and E. Ostroff (eds) *Universal Design Handbook.* New York: McGraw Hill, pp. 1.3–1.12.

Ostroff, E. and Iacofano, D. (1982) *Teaching Design for All People: The State of the Art.* Boston, MA: Adaptive Environments Center.

'You just follow the signs'
Understanding visitor wayfinding problems in the countryside

Katherine Southwell and Catherine Findlay

CHAPTER 8

One of the attractions of visiting countryside or woodland areas, for many people, is the opportunity to get away from familiar, and frequently urban, surroundings. Yet finding one's way in a new context is challenging and sometimes frustrating, particularly if the environment is remote and there are considerable distances involved. Wayfinding challenges can become a barrier, preventing people who are unfamiliar with the countryside from feeling confident about visiting new places. As government initiatives in Britain are developed to promote enjoyment of the countryside by a greater diversity of users (OPENspace, 2003), such issues as wayfinding assume a greater importance. This chapter discusses a project to explore certain problems associated with visitor wayfinding in forest and countryside recreation, and to develop tools to assist planners, designers and site managers in making investment in wayfinding infrastructure more effective. It was initially developed as a project for the Forestry Commission and therefore the examples relate to woodland sites. However, the work has since been demonstrated to be equally relevant in a wider range of countryside contexts, including the Peak District National Park and Durham Heritage Coast (both in England), and there is potential application in a wider, international context.

The starting point for this research was concern over the wide variation in visitors' satisfaction with road signs guiding them to different Forestry Commission sites in Britain. The visitor survey ratings ranged from 38% to 100% satisfaction, depending on the site, suggesting this might be concealing a broader problem in relation to accessing countryside facilities, and that road signs and wayfinding in general needed to be reviewed.

In an initial scoping study, the research explored visitors' wayfinding issues and confirmed the nature of the problem. Subsequent phases used a closer examination of the difficulties countryside visitors face to develop a set of guiding principles for good wayfinding design and evaluation. Although the initial scope of the research was to study the effectiveness of visitor road signs, the research examined signage in the wider context of wayfinding by exploring the full range of information visitors use from the start of their journey, for example tourist information in leaflets and on websites, and included analysis of the physical landscape as experienced in the view from the roadside. Subsequently, a set of principles and procedures became consolidated in the 'Site Finder' toolkit – a proprietary methodology for assessing visitor information and wayfinding needs in outdoor recreation (Southwell et al., 2007).

Since the overall aim of the research was to develop a problem analysis toolkit, the study findings were less concerned with identifying the wayfinding difficulties for visitors to particular sites, than with refining the techniques to analyse these problems on any site (Ward Thompson et al., 2005). A key innovation of the research was in its conceptualization of the data findings and the way in which the data were used to model the wayfinding experience in general for forest and countryside recreational visitors. This provided a contextualizing framework and organizational structure for understanding wayfinding in the outdoor (and often natural) landscape settings, as opposed to the built context.

Wayfinding and visitors to the countryside

Wayfinding systems for countryside recreation include all the sources of information that visitors use to find sites and enhance their recreational experience. Signs are a major element of investment in such systems because they are 'the most visible manifestation of corporate face' and 'provide reliable and accessible information to encourage and welcome visitors' (Forest Authority, 1997). Good signs also form part of a positive perception of woodlands and countryside areas and must be considered within the context of removing barriers to the countryside for people with disabilities and other socially excluded groups (Countryside Agency, 1998; Burgess, 1995; Ward Thompson et al., 2004).

The role of wayfinding signage is essentially to eliminate the unknowns in navigation (Caves and Pickard, 2001). Evidently, signage plays a lesser role for a person who knows an area well, but to a first time visitor, particularly in spatially complex environments, people may be totally reliant on signs for navigation, especially in the absence of good maps or good skills in map reading. The unknowns in wayfinding create uncertainty and frustration and increase levels of anxiety and stress. This has a negative effect on the visitor experience, with a particular impact on the leisure-seeking aspect of outdoor recreation which is concerned with relaxation and enjoyment.

One of the main difficulties highlighted by the Forestry Commission study is that much of the research on wayfinding has taken place in the context of the built environment and indoor places such as airports and hospitals. There is consequently a lack of understanding about whether these principles apply equally to outdoor settings. In addition, there is a lack of practical knowledge about how wayfinding understandings relating to the wider outdoor environment may be used effectively for environmental planning and design purposes. It seems that if design is to play a role in improving wayfinding ease in a place, the wayfinding activity needs to be understood in terms of the physical and operational environments in which it occurs (Carpman and Grant, 2002). However, a fundamental characteristic of environmental wayfinding is that it is both a physical activity and a psychological process.

The fundamentals of wayfinding

Wayfinding is 'the ability to identify one's location and arrive at destinations in the environment, both cognitively and behaviourally' (Prestopnik and Roskos-Ewoldsen, 2000) or, more simply, 'spatial problem-solving' (Passini, 1992, 1996). It is a dynamic cognitive process where movement through space requires a continuous involvement in reading, interpreting and representing that space (Appleyard et al., 1964). The wayfinding activity is also highly complex, involving a variety of search processes and sources of information, of which signs are one component, often supplemented by leaflets, published maps, personal contacts and word of mouth. An added complication is that wayfinding ability appears to differ between individuals depending on age, gender, sense of direction, familiarity with the environment and wayfinding strategy (Prestopnik and Roskos-Ewoldsen, 2000; Lawton, 1996; Lawton et al., 1996).

There is a general consensus that there are two principal strategies for wayfinding in humans: the first assumes an understanding of the spatial structure of the environment and key locations within it – a kind of mental map – where people rely on the spatial relationships between locations to navigate; the second is based on people's knowledge of places and the routes that connect them – the sequential experience – which are used to navigate. Users unfamiliar with an environment may start with one strategy and switch to another as they become familiar with a place.

Passini's conceptualization of the role of signage in wayfinding is useful: the physical setting and routes are seen as defining the problems that people will have to solve, while the graphic communication (the signs) provide the information to solve the problems. When involved in wayfinding in practice, people will pick up information from a number of sources, of which signs are a key element. The central problem for designers is to make the information provided by signs compatible with the information people are likely to pick up from the wider environment and compatible with their particular needs. Others concur with this contextualized view of wayfinding in the physical environment, recognizing that to achieve excellence in wayfinding design, we need to 'go beyond the sign' (Berger, 2005).

Getting wayfinding right in environmental design involves an ability on the part of the designer to imagine the whole experience in its entirety. This is because the human wayfinding activity is one like few others in that it demands a complete involvement with the environment (Passini, 1992). This author suggests that decisions are the key 'units' and structure for wayfinding and that this takes place in a decision hierarchy, starting with a general plan or goal 'to go to a place', which in turn translates into a series of tasks, for example 'head west', and sub-tasks which become ever more detailed on approaching a specific target destination, for example 'turn left, turn right'.

Landmarks certainly play a key role in wayfinding and help navigation of both familiar and unfamiliar territory (Lynch, 1960). Notably, Siegel and White's (1975) theory of navigation (wayfinding) conceptualizes a developmental progress starting with *landmark*, to *route*, to *survey* map. While it is not clear if each individual goes through these stages in such a logical sequence, it provides a framework on which we are able to build a mental representation of the physical environment. It is notable that landmarks are only useful if we know something about the relationship between the landmark and our wayfinding goal (i.e. if we understand what they are and where they are) and so are unhelpful to visitors who are new to a place, although, as highlighted by Jakle (1987), landmarks and views can be useful to people who have seen images of them or had them described prior to their visit.

There is a complex array of external (environmental) information to process when we are moving through an environment. Downs and Stea (1977) suggest that our information processing for wayfinding is broken down into discrete stages which include orientation, route decision, route monitoring and destination recognition. Passini (1996) notes how selective we become in our information processing when moving through an environment, and in particular when moving at speed, for example driving. A sign or other feature in the landscape might not register in our minds because we are not actively seeking it out or because we are not expecting to see it. This selectivity in information processing when moving through the landscape can have important consequences for the design and safety of road systems on approach to a forest/countryside recreation site, as well as the design of entrances and arrival areas on site, in particular for helping to understand what the minimum visitor information requirements are.

Visitor information needs in countryside recreation

Visitor surveys carried out by the Forestry Commission indicate that most people arrive by car. While it is recognized that it is not acceptable for sites to be designed around the motor vehicle alone, this remains the only means of access to many countryside sites for the majority of people at present (Findlay *et al.*, 2002) and so the research has focused particularly, but not exclusively, on this mode of transport.

Existing signage systems at UK forest and countryside recreation sites are typically structured using an information hierarchy as follows: on approach to site, pre-arrival signs take the form of roadside warnings such as 'brown' (dedicated to tourism) and other highway signs, while threshold signs announce a special area has been arrived at, and/or raise awareness of the organization or landowner responsible for managing the site (Bell, 1997; Forest Enterprise, 1998; Winter 1998). Once on site, the visitors' need for orientation and directional signs is provided for, with some further information for identification and additional information, interpretation and regulation as appropriate.

One of the key gaps in knowledge highlighted by our research related to the practical question of how to measure, and therefore evaluate, the performance of signs in outdoor wayfinding systems. A literature search of post-occupancy evaluations of airports, hospitals, shopping malls and other areas of the built environment revealed a potential set of criteria including visual accessibility, consistency in directional and reassurance information, route connectivity (or ease of movement through a route system from one intersection to the next), clarity of message and legibility of site layout, as some of the criteria for good wayfinding design, but it was not clear how consistently these might be applied to the outdoor landscape experience. It became apparent that there was much to learn about the specifics of outdoor wayfinding in the sphere of countryside recreation.

Methodology used in the wayfinding study

A multi-method approach was used in a research methodology that was piloted during the first phase of the research, using five case study sites across the United Kingdom. This was further developed and refined during a second phase of work, using a further four sites. The methodology comprised a combination of qualitative environmental survey and interview techniques. The latter comprised semi-structured interviews, while the former used environment–behaviour observation, role-play, visual techniques for spatial analysis and a photographic sign audit. Key approach routes to sites, entrances, and on-site arrival areas were examined. The basic questions to address were: are visitors finding their way easily to the sites they wish to visit?; does the information provided on site enable visitors to use the site effectively once they have arrived?; and is the issue to do with signage or some other factor?

The nine case study sites are listed in Table 8.1. These varied in size and type, ranging from a local amenity urban fringe woodland site to a high-profile major visitor destination site with national (and international) visitor catchment.

Interview survey techniques

During the first phase of the research (the scoping study), semi-structured interviews were used with open-ended questions. The interviews were divided into sections, designed to follow the sequence of arriving and spending time on the site (Findlay et al., 2002). In the early stages of the study, the effectiveness of the sequential approach for capturing the details of the visitor wayfinding experience quickly became apparent. However, it was also found that interview questions had to be carefully phrased in order to elicit vital information. For example, when visitors were asked if they had experienced any problems finding a particular site, frequently people would initially state that they had no problems, but then later reveal that in fact they had had some kind of difficulty.

The critical technique for eliciting the 'hidden' information from people was to encourage individuals to tell the story of their journey in detail, one stage at a time, rather than progress a line of questioning about their own wayfinding strategies and abilities. By reliving the journey in sequence, one stage at a time, from the point of leaving home, or other point at which the decision was made to make their visit (this could have been a last-minute decision from the roadside) through to arrival on site, visitors could focus on the environmental factors and identify the key sources of information and signs which played a role – where and at what point in the journey.

The combined use of interviews, role-play and observations provided an effective mechanism for identifying the full range of factors that structured visitors' decision making, and so help understand the fuller scope of 'the wayfinding problem'. Comparisons across different sites helped distinguish 'good' wayfinding systems from 'bad' ones. These were identified through interview responses, where certain indicators of a successful signage system emerged. For instance, a site that visitors found easy to find through a complex maze of country roads would be indicated by the response 'I just followed the signs'. This contrasted with a visitor's description of a difficult-to-find place which was correspondingly more difficult to describe: 'take M4, Junction 40, brown sign then go through two roundabouts, up valley, past big rock . . . travel on a bit and turn right into the site'. Thus, a key indicator turned out to be that an easy route to use was also an easy route to describe. While these responses were effective for highlighting a potential problem and/or focus area for follow up analysis, interviews alone could not identify the specific nature of a problem. For this, the environmental survey techniques were needed to examine the issues in physical context.

VISITOR WAYFINDING PROBLEMS IN THE COUNTRYSIDE

Table 8.1 Phase I and II case study sites

Site	Location	Site typology	Size (ha)	Visitor nos per annum	Main Use	No. of interviews
Queen Elizabeth Forest Park	Trossachs, Scotland	Regional, National (International)	20,000	1,000,000	Walking, Cycling	22
Glencoe Lochan	Scottish Highlands	Regional, National	137	30,000	Walking, Fishing	21
Cannock Chase	Midlands, England	Local	2,428	106,000	Walking, Cycling, Horseriding	9
Dalby Forest	Yorkshire, England	Regional	3,642	300,000	Walking, Cycling, Mountain biking	12
Hafren	Mid Wales	Local	3,000	20,000	Walking	6
Afan Argoed	South Wales	Regional	3,250	150,000	Mountain biking, Mining museum	27
Coed-y-Brenin	North Wales	Regional, National, (International)	3,600	100,000	Mountain biking	27
Pages Wood	Essex, England	Local	75	n/a	Community forest, Walking/cycling	26
Moors Valley	Dorset, England	Regional	300	800,000	Family day out	29

Environmental survey techniques

The combined use of methods (described below) in role-play, spatial analysis, observations and the sign audit not only provided a contextualizing mechanism, but also facilitated a comparison between what information people said they saw and/or expected to see, and what was actually provided in the signs or physical environment.

a) Role-play and spatial analysis

Role-play was particularly useful for analysing issues that were difficult to observe in people, for example their experience when driving along the road. Role play was used combined with a spatial analysis technique frequently used in landscape architecture for visualizing movement through a landscape as a sequence of images (commonly termed 'spatial sequence technique'). This can be done either at walking speed or at driving speed to explore the view from the road. The images could be put together in sequence as a 'slideshow' for a subsequent desk study to re-examine the route experience as part of the study analysis. This helped elicit all the subtle details of the landscape that might have played a vital role in the wayfinding experience but might not be immediately obvious.

8.1 Observing visitors' driver behaviour at a site entrance.

b) Observations

It was useful to observe visitors' behaviour for how they responded to signage, and how their behaviour appeared at times to contradict the version of events relayed in the interviews. For example, an interview survey conducted at a site in Dorset, England gave no indication of there being any problems with the entrance layout, or its signage, and yet driver behaviour indicated otherwise: on approaching the entrance, visitors would typically brake suddenly at the entrance, indicator lights would start very late in the manoeuvre, and more often than not visitors would make a sudden swerve in order to enter the site (Figure 8.1).

Once parked and out of their cars, visitors were observed on site trying to find key facilities needed on arrival, such as the toilets. Peoples' body language, gestures and verbal exclamations were noted and used as indicators of problems requiring further study. Trying to identify the key cause of the problem inevitably led to a realization of what the solutions might be. For example, at a site in mid Wales, a sign to the toilets from the car park pointed along a path whose end was not in view; when the toilet block was not clearly visible to people after 20 or 30 metres of walking along the path, visitors began to back track their route, doubting the information they had seen on the sign. Further analysis (through role-playing the experience) revealed that the toilets would have been much easier for people to find if the trees had been cut back to show the toilet block from the car park.

c) Sign audit

Whereas role play, spatial analysis and behavioural observations captured the experiential aspects of wayfinding, the sign audit provided a factual survey of graphical information (in signs). For this, the routes and entrances were photographed at walking speed and a complete inventory of signs generated.

Results

The initial scoping study established the existence of a wayfinding problem. The study indicated a general lack of fit between information provision and visitor information requirements as the key causative factor, and that signage can be assessed in relation to three inter-related aspects of wayfinding summarized as a 'people, purpose, place' model illustrated in Figure 8.2.

The 'people, purpose, place' model highlights the role of signs in the wider context of wayfinding: essentially, signs need to be directed at the relevant user group (i.e. 'people'); information should respond to what the visitor needs to know most at each stage of the journey (i.e. have 'purpose'); and that information should be positioned where the wayfinding message can have maximum effect in its physical context (i.e. respond to 'place').

The second phase of the wayfinding study sought to identify the specific attributes of 'the wayfinding problem' in countryside recreation, and distinguish between visitor problems that were specific to a given site and those that were generic. It also sought to distinguish between problems that relate to intrinsic site factors that cannot easily be changed, for example a complex road system in that part of the country, and those that can readily be changed, for example a hidden entrance sign. Results overall highlighted that most of the wayfinding difficulties experienced by first-time or infrequent visitors can be attributed to one of four problem categories:

1. **Inconsistency in names and labels used for the site:** Visitors to a site will look out for signs, symbols, names, landmarks and so on that are named and indicated on the initial information source they access, for example a website or tourist information leaflet. If the names, symbols and appearance of information subsequently encountered are different, people become uncertain or confused. The more consistent information is, regardless of its source, the more confident people will be in using it to find the site.

2. **Lack of advance warning and reassurance at route junctions:** Visitors approaching a route junction need both an advance warning sign or signs, directing them if and where they should turn off, and reassurance after the junction that they are still on the correct route. Reassurance can be provided by landmarks or signs. Both are particularly important for road drivers.

3. **Missing the site entrance:** Drivers approaching a site will normally prepare to turn off the road only when they see both a sign marking the entrance and an obvious entrance gap in the roadside. A hidden road entrance and/or an obscured or nonexistent entrance sign means many visitors will drive past the site before they have time to react, or they will brake suddenly (and potentially dangerously) in order to make the turnoff. Although missing the site entrance is a particular hazard for drivers of motor vehicles, it can also be a problem for horse-riders, cyclists and even pedestrians, if the entrance is obscure.

4. **Becoming confused, lost or frustrated once on-site:** When visitors arrive, they want to be reassured that they are in the right place and find out where certain facilities are quickly. Drivers want to know where to park and where/if they have to pay, many people want general information about facilities such as toilets and information centres and some visitors want to go straight to key activities such as mountain biking.

Results suggested a sequence of visitor information needs that fall into the four distinct categories suggested above, each of which are decisional stages important in wayfinding (Ward Thompson et al., 2005). This suggested a basic framework for wayfinding analysis with potential application for structuring a practical toolkit. However, translating this from a basic framework into a practical set of evaluation tools and techniques presented a significant challenge, requiring a degree of innovation: methods had to be devised that would be simple enough for site managers with no specialist research expertise to use,

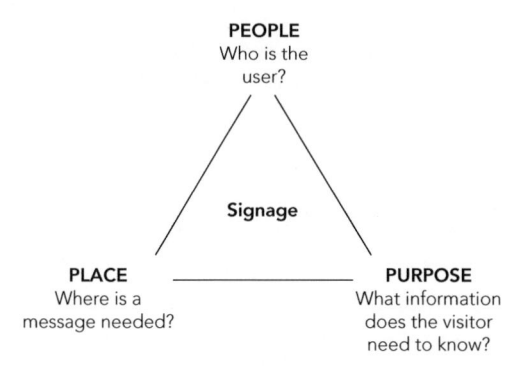

8.2 A wayfinding signage model.

PEOPLE
Who is the user?

Signage

PLACE
Where is a message needed?

PURPOSE
What information does the visitor need to know?

and yet provide effective problem analysis within a robust site evaluation process.

Developing the 'Site Finder' toolkit

The toolkit was specifically designed with the site manager in mind. This required tools that would help them to see their sites through the eyes of the first time visitor, with 'fresh eyes', to understand where visitor information and wayfinding problems lie, and what changes might make a difference. There is a particular focus on signage evaluation, but the toolkit is designed to assess them within the context of the whole wayfinding experience.

The toolkit comprises a two-stage process of survey and analysis (Table 8.2). There are three survey tools and four analysis tools, each of which relates to a wayfinding information sequence, based on the four problem categories described above.

As indicated in Table 8.2, there is a critical 'need to know' question arising at each stage of the journey. Thus, at pre-arrival stage, the visitor needs to know 'what is the site called?'; along the road he/she needs to know 'how do I get there?'; and then on approaching the site 'where is the entrance?'; and, finally, on arrival when needing on-site orientation, the visitor needs to know 'what to do/where to go?'.

The three site survey tools are then used to screen for potential problems, to establish if problems exist in relation to the four decisional stages (i.e. pre-arrival, the route, the entrance, on-site arrival), and thus where further analysis should focus. There are four analysis tools for each of the categories.

It is only possible here to briefly outline some of the key elements of the toolkit. The full toolkit is described in Findlay et al. (2003) and the final version, named the Site Finder wayfinding analysis toolkit, will be published in a form for practitioners and end-users in 2007. While the three survey tools are a vital element of the toolkit, only the four analysis tools will be described here.

Table 8.2 The wayfinding information sequence: a framework for survey and analysis

Problem category	SURVEY (three site survey tools)	Focus area	ANALYSIS (four tools) Criteria for assessment	Method
(1) Pre-arrival information: *What is the site called?*	a) Information survey	Site name	**i) Informational consistency**	'information sign-line'
(2) Route information: *How do I get there?*	b) Visitor Questionnaire	Key road junctions on approach route/s	**ii) Route connectivity** (or ease of movement through key route junctions)	'decision tracking'
(3) Finding the entrance: *Where is the entrance?*		Entrance	**iii) Entrance reassurance**	'triangle of vision'
(4) Finding out what to do and where to go on arrival. *How can I find out what to do next?*	c) Sign survey	Arrival area on site	**iv) Site legibility** (or ease of understanding of site layout)	'behaviour mapping'

VISITOR WAYFINDING PROBLEMS IN THE COUNTRYSIDE

Analysis tools

The four analysis tools provided in the toolkit comprise four criteria for assessment with a corresponding analytical method for each. The criteria are the 'measures' used to assess signage for wayfinding performance. Thus, for tool no. 1 *informational consistency* is measured using an 'information sign-line'; tool no. 2 measures *route connectivity* using a 'decision tracking' method; tool no. 3 measures *entrance reassurance* using a 'triangle of vision'; and, finally, tool no. 4 measures *site legibility* using a 'behaviour mapping' method.

The purpose of the analysis is to focus on how a site is seen through the eyes of a first-time visitor, and to fully experience the visit with 'fresh eyes'. For this, each of the tools incorporates an element of role-play.

Tool no. 1: Informational consistency

This tool assesses informational consistency throughout the journey from pre-arrival, along the *key* route (i.e. the most obvious route choice from the perspective of a first-time visitor) and on arrival at the entrance. In particular it assesses how information provision in the *sign content* relates to other pre-arrival sources of information in circulation in print (e.g. in leaflets advertising the site or on road maps), on the Internet and by word of mouth.

Rationale

The surveys indicated that the initial source of information about a site name dictated what information visitors were seeking along the road and at the entrance – from point of departure to arrival – and lack of consistency in names used as well as variation in the style of signs caused visitors much confusion.

Application

The various names used across all pre-arrival sources of information, all signs along the roads and at the entrance are written out, for example on 'post-its'. At a good (or easy to find) site there is only one line of information, since the same name is consistently used in leaflets, in signs along the road and for entrance signs. Figure 8.3 illustrates an example of a site with good informational consistency, with just one point where the information line deviates, or branches off. By contrast, at a site with many inconsistencies in the way the site is referred to in leaflets and signs, there will be a many-branched line as multiple lines of information are introduced. Thus, a high number of branching points in the 'sign-line' would rate low on consistency. The technique for this is termed 'information sign-line'.

Even if the sources of information on access to the site are highly consistent, and the signage content itself clear and consistent on the route to site, the visitor may still have trouble finding a place because of poorly positioned signs at road

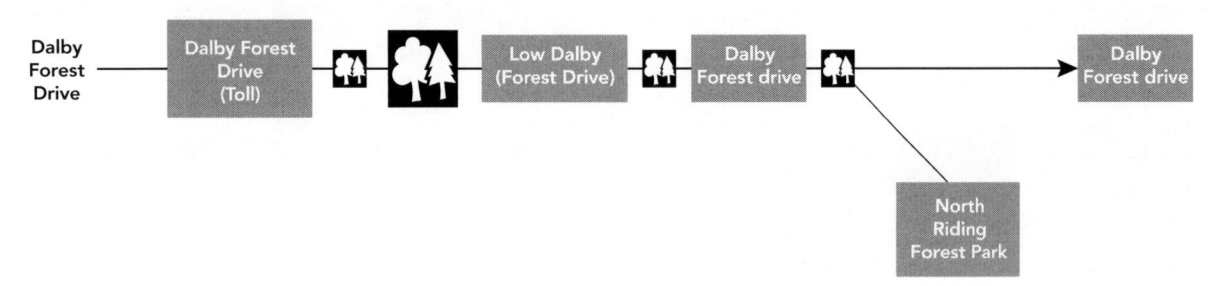

8.3 An 'information sign-line': the low number of branches indicates a site with good informational consistency.

KATHERINE SOUTHWELL AND CATHERINE FINDLAY

junctions. While tool no. 1 focuses on sign *content*, tool no. 2 examines sign *placement*.

Tool no. 2: Route connectivity

This tool examines the effectiveness of signs at road junctions. The tool assesses if and how signage at key junction points on key feeder routes connects people with the local route network, in the *key* direction(s) of travel.

Rationale

Visitor confusion occurs where key decision points are not signed or the signage is obscured or not legible at road junctions.

Application

The tool uses a technique termed 'decision tracking'. This entails breaking down, step by step, the experience of turning off at a junction, where a problem is suspected. Thus the driving experience is 're-lived' at slow driving speed or walking pace (if safe to do so) to examine in detail all the informational cues at a junction, to try and pinpoint the precise nature of the problem. For example, Figure 8.4 illustrates a road junction with a critical wayfinding sign that visitors frequently missed at driving speed: on close examination it was found that there was a sign giving clear direction to the site, but it was hidden in a clutter of other signs.

The tool helps identify: a) problem spots along the route; b) if the problem relates to a lack of signage – directional and/or reassurance signage; c) whether problems are directly related to lack of sign visibility – whether it is sign clutter, tree growth or misalignment.

Tool no. 3: Entrance reassurance

This tool assesses visual accessibility to the entrance and entrance signage. It examines the effectiveness of the visual cues at an entrance that signal 'arrival'.

Rationale

When comparing 'good' and 'bad' entrances, as indicated in both visitor interviews and through observation of visitor behaviour at entrances, it was found that the best sites indicated advance warning of their entrance in combination with a clear view of the entrance. In the worst example, visitors frequently drove straight past because the entrance confirmation sign appeared at exactly the same moment as drivers would be passing the entrance, situated as it was around a sharp bend in the road. By the time visitors had made a visual connection between sign and entrance, it was too late to brake. In the observed example, the visitors were able to turn around and come back, but this is not always easy to do.

It did not seem to be important whether the distance to the entrance was indicated or not on the advance warning, since visitors did not appear to respond until they could see the entrance opening itself – something that had to appear 'entrance-like'.

When two models of 'good' and 'bad' entrance were used to plot the point at which both entrance and sign came into view, it was found that a triangle of vision could be drawn between the driver's position, the sign and the entrance opening: the more obtuse the angle of the triangle, the shorter the space/time frame the user is provided with in order to react, while a more acute triangle indicated a longer view of the entrance/sign arrangement in the road ahead – as illustrated in Figure 8.5.

8.4 Sign clutter at a road junction.

Application

Application of the tool involves a field survey technique termed 'triangle of vision' (Figure 8.5). The points of vision are plotted by role-playing the driving experience and drawing a triangle between the point at which the entrance and sign together come into view at the normal speed of travel of the road. This requires role-playing the driving experience along the road with 'fresh eyes', responding to the visual cues as if visiting for the first time.

Tool no. 3 helps establish if and what 'type' of entrance problem the site has:

- whether the entrance opening and/or entrance signs are obscured because of the layout of the road and other intrinsic environmental factors (that cannot be changed);
- whether the entrance opening and/or entrance signs are obscured because of a lack of maintenance such as cutting back trees or other factors that can be changed;
- whether the entrance/sign layout needs to be reconsidered;
- if and where a new sign is needed.

Tool no. 4: Site legibility

This tool helps identify problems related to *spatial legibility* in the arrival area. It analyses visual accessibility to the key facilities, and/or information for helping users locate these.

Rationale

Immediately upon arrival, visitors need to know *where* key facilities are located in *relation* to the arrival point. This implies the need for clear lines of visibility across a site from key arrival points, to enable visitors to create a 'mental map' of the route from the car park to, for example, the visitor centre, information point, toilets or start of walk.

In respect of these arrival needs, the on-site signs found in the survey were not always helpful. There was a general lack of consideration of site layout, in particular lines of visibility across site, resulting in directional and reassurance signs seemingly

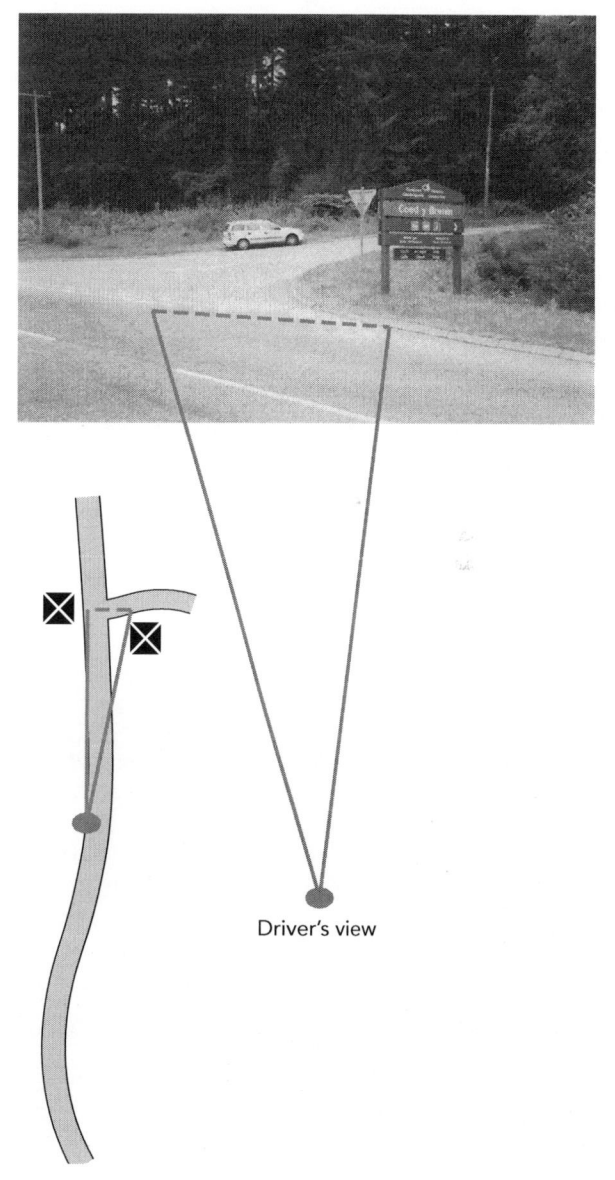

8.5 A 'triangle of vision' technique for analysing the entrance.

pointing at nothing in particular. Failure to understand the layout of site on arrival could limit the user's ability to use the site. This can have implications in terms of whether the visitor uses the facilities at all, and visitor satisfaction overall.

Typical 'need to know' issues subsumed within the general issue of 'what to do/where to go', were:

a) Have I arrived at the right place?
b) Where is the car park?
c) Where do I pay/do I have to pay?
d) Where is the visitor centre/key information point?
e) Where are the toilets/are there any toilets?
f) Where does the key activity start? (e.g. mining museum, start of the trails, café or picnic area).

Application

The approach involves analysing the visitor experience through a combination of observation and role-play of specific tasks, for example 'find the toilets from the car park'. Visitors' routes across a site are plotted to examine how far people have to walk to gain visual reassurance of where a facility is located. The principle is that the longer the route, the less legible the site is.

Tool no. 4 can achieve the following:

a) pinpoint problem spots around the arrival areas;
b) identify if the problem relates to a lack of signage (directional and/or reassurance signs) – or signage surplus, or signs saying nothing helpful;
c) identify whether problems are directly related to lack of sign visibility – whether it is sign clutter, tree growth or bad positioning;
d) identify lines of visibility across a site.

Discussion

The analytical process which was followed through the research investigation and subsequent translation into a toolkit demonstrated the value of the mixed method approach. Wayfinding response actions could be examined in detail using a combination of role-play, observation and appropriate visitor interview, together with visual techniques in spatial analysis to identify the specific characteristics of wayfinding particular to outdoor countryside recreation. Choosing an appropriate method for analysis of each of the key aspects of wayfinding helped sharpen understanding of the whole process. The resultant research product (the toolkit) has thus provided a vehicle for translating theory into practice.

The critical mechanism is the four-part wayfinding sequence structure (Table 8.2). This facilitates the analytical process and demonstrates a decision-based model of outdoor wayfinding for universal application in countryside recreation. The use of decisions as the units and structure for wayfinding evaluation is key to the whole process: these help the assessor to step inside the mind of the first-time visitor and re-live the journey in a methodical way. In so doing, the unique attributes of each given situation are analysed. Although we cannot 'see' wayfinding decisions we can see people's actions directly (through observation) or indirectly (through role-play and interviews) and relate the environmental information (whether a sign or a view across a site) to these response actions. This was the basis on which each of the analytical tools were developed within the Site Finder toolkit.

The toolkit draws on a combination of survey and analytical methods for problem identification and subsequent problem analysis. The process has been developed as a systemized but highly flexible approach to wayfinding evaluation. Thus, if a site is intended to remain a highly localized amenity and not attract a wider visitor catchment, there is no requirement to assess the site beyond the entrance. In this instance the evaluation process can focus on entrance reassurance and site legibility issues only.

The two-layered approach to wayfinding analysis which first identifies a problem and then examines it in detail, is vital to the process: without in-depth analysis of a given (or suspected) problem there can be no understanding of the specific characteristics of each given situation, and subsequently what the potential solution types might be. For example, the 'triangle of vision' technique forces a detailed look at each single component of the whole entrance experience as if in slow motion, so pinpointing the specific causation factors and therefore what to do about the problem, for example cut back overgrown trees or move a sign.

The mixed method approach used in the study was dependent on visualization techniques to analyse the spatial problem-solving aspect of wayfinding. In other words, the toolkit forces a look at a wayfinding system in which the background environment setting is seen as an integral part of visitor information processing, for example making visual links between sign and entrance-like opening on the roadside, and which helps one appreciate the array of environmental messages that operate 'silently' in the landscape. Car park layouts, for example, can be designed to make it so obvious to visitors where they have to go next that there is no need for signs. However, at a number of sites surveyed in the research, signs were frequently placed unnecessarily, so adding clutter and potentially further confusion. In fact too many signs or confusing signs can be as much of a problem as too few (Ward Thompson et al., 2005) and in many forest and countryside parks signs have been added at different times in an *ad hoc* fashion, so increasing the possibility for inconsistencies in information. This would suggest the need for wayfinding evaluation to help review, reorganize or rationalize signs in a wayfinding system, rather than keep adding signs to address individual problems. This would also suggest that management of views into and across a site should be seen as a vital component for outdoor wayfinding design.

Conclusion

The process and practice of wayfinding design in countryside recreation must be based on the type of navigational problems people encounter in the outdoor and natural (as opposed to built) context. To this end, the research has provided some new conceptualizations of environmental wayfinding in the context of countryside recreation. Significantly, the research process and its product (the Site Finder toolkit) have given form and structure to wayfinding evaluation in outdoor recreation where none existed before, providing a vehicle for translating theory into practice, as well as a framework for future wayfinding studies.

The key innovation of the toolkit is its deconstruction of the wayfinding problem into a simplistic, and yet holistic, four-part sequence. The sequence captures the logic of wayfinding as a hierarchy of decision making from pre-arrival, along the routes, the entrance and arrival areas on-site, so providing the basic building blocks for wayfinding design and evaluation.

Overall, the research presents a robust framework for wayfinding analysis developed for use by forest, woodland and countryside recreation site and park managers. This enabled the Site Finder toolkit to take into consideration the full spectrum of information that visitors use – ranging from visitor leaflets, signs along the road, entrance cues and information panels – helping to see a site's problems through the eyes of first-time visitors. In allowing the users' perspective to drive the process of problem analysis and identification of solutions, this provides a practical way of determining priorities for effective investment.

Acknowledgements

This research was undertaken between 2000 and 2004 by OPENspace research centre at Edinburgh College of Art/Heriot-Watt University on behalf of the Forestry Commission.

References

Appleyard, D., Lynch, K. and Myer, J.R. (1964) *The View from the Road*. Cambridge, MA: MIT Press.

Bell, S. (1997) *Design for Outdoor Recreation*. London: E & F Spon.

Berger, C.M. (2005) *Wayfinding. Designing and Implementing Graphic Navigational Systems*. Mies, Switzerland: RotoVision.

Burgess, J. (1995) *Growing in Confidence: Understanding People's Perceptions of Urban Fringe Woodlands in Northampton*. Northampton: Countryside Commission.

Caves, R.E. and Pickard, C.D. (2001) 'The satisfaction of human needs in airport passenger terminals', in *Transport*, vol. 147 (1), pp. 9–15.

Carpman, J.R. and Grant, M.A. (2002) 'Way finding: a broad view', in R.B. Bechtel and A. Churchman (eds) *Handbook of Environmental Psychology*. New York: John Wiley and Sons, pp. 427–442.

Countryside Agency (1998) 'Barriers to enjoying the countryside'. *Research Note CCRN 11*. The Countryside Agency.

Downs, R.M. and Stea, D. (1977) *Maps in Minds: Reflections on Cognitive Mapping*. New York: Harper & Row.

Findlay, C. and Southwell, K. (2004) 'Visitor information and wayfinding needs', in M. Bull (ed.) *Visitor Information and Wayfinding Needs. Seminar Proceedings of the Countryside Recreation Network*. Sheffield: Countryside Recreation Network.

Findlay, C., Southwell, K., Ward Thompson, C. and Aspinall, P. (2002) 'The effectiveness of wayfinding systems with forest users', *Conference Proceeding: Monitoring and Management of Visitor Flows in Recreational and Protected Areas*, February 2002, Vienna.

Findlay, C., Southwell, K., Ward Thompson, C. and Aspinall, P. (2003) *Effectiveness of Wayfinding Systems with Forest Users*. Phase 2 Report to the Forestry Commission, Edinburgh College of Art/OPENspace unpublished report.

Forest Authority (1997) *Woodland Signs: Good Practice*. Edinburgh: The Forest Authority, Forestry Commission.

Forest Enterprise (1998) *Forest Signs*. Forest Management Memorandum 27. Edinburgh: Forestry Commission.

Jakle, John A. (1987) *The Visual Elements of Landscape*. Amherst: The University of Massachusetts Press.

Lawton, C.A. (1996) 'Strategies for indoor wayfinding: the role of orientation', in *Journal of Environmental Psychology*, vol. 16, pp.137–145.

Lawton, C.A., Chartleston, S.I. and Zieles, A.S. (1996) 'Individual and gender-related differences in indoor wayfinding', in *Environment and Behaviour*, vol. 28 (2), pp. 204–219.

Lynch, K. (1960) *The Image of The City*. Massachusetts, MA: The MIT Press.

OPENspace (2003) *Diversity Review – Options for Implementation*. Prepared for the Countryside Agency by OPENspace Research Centre, Edinburgh. Available at www.countryside.gov.uk/LAR/Recreation/DR/Scoping.asp.

Passini, R. (1992) *Wayfinding in Architecture*. New York: Van Nostrand Reinhold.

Passini, R. (1996) 'Wayfinding design: logic, application and some thoughts on universality', in *Design Studies*, vol. 17, pp. 319–331.

Prestopnik, J.L. and Roskos-Ewoldsen, B. (2000) 'The relations among wayfinding strategy use, sense of direction, sex, familiarity and wayfinding ability', in *Journal of Environmental Psychology*, vol. 20 (2), pp. 177–191.

Siegel, A. and White, S. (1975) 'The development of spatial representations of large-scale environments', in *Academic Press*, vol. 10, pp. 9–55.

Southwell, K., Ward Thompson, C. and Findlay, C. (2007) *Site Finder: assessing the countryside visitor's wayfinding experience*, Edinburgh: OPENspace Research Centre.

Ward Thompson, C. (2004) 'OPENspace Research Centre and countryside visitor needs', in M. Bull (ed.) *Visitor Information and Wayfinding Needs. Seminar Proceedings of the Countryside Recreation Network*. Sheffield: Countryside Recreation Network.

Ward Thompson, C., Findlay, C. and Southwell, K. (2005) 'Lost in the countryside: developing a toolkit to address wayfinding problems', in B. Martens and G. Keul Alexander (eds) *Designing Social Innovation: Planning, Building, Evaluating*. Göttingen: Hogrefe & Huber Publishers, pp. 38–45.

Ward Thompson, C., Aspinall, P., Bell, S. and Findlay, C. (2004) *Open Space and Social Inclusion: Local Woodland Use In Central Scotland*. Edinburgh: Forestry Commission.

Winter, P. (1998) *Getting out the message . . . an examination of signs in recreational settings*. US Forest service – Recreation Research Update. October 1998. No. 28.

Developing evidence-based design

Environmental interventions for healthy development of young children in the outdoors

Nilda G. Cosco

CHAPTER 9

The sedentary lifestyle problem

It is well established that behavioural patterns such as food intake, physical activity and sedentary lifestyles have a strong impact on childhood obesity (Davison and Birch, 2002). This chapter is a summary of the background for a study of childcare outdoor play environments conducted by the author as a research contribution to the emerging field of design for active living for young children. The study is based on the positive association between physical activity and weight in children (Sallis *et al.*, 2000), the premise that the childcare centre is the highest predictor of physical activity of children 3–5 years old (Finn *et al.*, 2002) and the notion that the childcare outdoors is the strongest correlate of physical activity of preschool children (Baranowski *et al.*, 2000; Sallis *et al.*, 1993).

Most young children learn about the surrounding world by physically interacting with it. For them, life is movement and sensory stimulation (Piaget, 1952). The neural pathways of the brain are developed through movement, revealing a clear interdependence between physical activity, language acquisition and academic performance (Hannaford, 1995). Play is the motivating force that produces physical activity (Pellegrini *et al.*, 1998) and social interactions with other children and adults (Frost *et al.*, 2001; Moore and Wong, 1997). Despite these assumptions about children's natural drive to stay active, the health of even the youngest is currently affected by sedentary lifestyles. Recent research in young Scottish children by Reilly *et al.* (2004) shows that 3-year-old boys spend a median of 76% of their time in sedentary behaviour and girls up to 81%. The proportion reduces slightly with age. Time spent in sedentary behaviours for 5-year-old boys shows a median of 73%. For girls the median is 78% (Reilly *et al.*, 2004). This lack of physical activity, combined with poor nutrition, is producing a profoundly negative effect on children's physical health, especially in developed countries. In the United States, more than 10% of children two to five years old are overweight and more than 20% of children of the same age are at risk of being overweight (Ogden *et al.*, 2002). The average rate of child obesity in Europe is 25%, with highest percentages in Spain (30%) and Italy (36%) (IOTF, 2005). In the United Kingdom, 16% of children aged two to fifteen are obese according to the Health Survey for England (2002).

These figures imply serious health problems, especially for low-income children (Mei et al., 1998), and suggest the need for urgent interventions if children are to avoid a compromised quality of life at an early age. The added cost associated with such dramatic decline in health is also an issue.

Daily life environments for many young children

According to the National Survey of America's Families, in 1999 almost three-quarters (73%) of US children under five with employed parents were in a childcare arrangement other than care by a parent (Capizzano et al., 2000; Sonenstein et al., 2002). This percentage represents approximately 8.7 million children from which 42% (3.6 million children) spend most of their waking hours in centre-based and family childcare services (28% and 14% respectively).

In effect, in the last two decades, childcare centres in developed countries have become the most crucial environment outside the home for young children. Despite this fact, researchers and governmental agencies have been slow to consider childcare centres (highly regulated institutions) as gateways for environmental interventions aimed at obesity prevention early in life.

For many years, even the most influential health reports avoided reference to young children. The *US Surgeon General's Report on Physical Activity and Health At-a-Glance Summary* (1996: 1) 'brings together, for the first time, what has been learned about physical activity and health from decades of research'. The publication includes vital information about a new view of physical activity and its benefits, a call for moderate exercise in daily life, precautions for a healthy start at different stages of life and special messages and guidance for different population groups. Remarkably, children under 12 years of age are not mentioned.

Just recently, on 23 April 2003, in a speech to the National Head Start Health Institute, Washington, DC, Vice Admiral Richard H. Carmona (then US Surgeon General) called childhood obesity 'the most serious health problem in America today'. Such forthright concern should make the topic a US national priority with an implied call for action for educators, health professionals and licensing consultants to produce the appropriate institutional changes to address this health crisis.

The built environment and children's active living

In 2004, the US National Institute of Health (NIH) acknowledged the impact of the built environment on sedentary lifestyles in the Strategic Plan for NIH Obesity Research entitled *Preventing and Treating Obesity through Behavioral and Environmental Approaches to Modify Lifestyle* (USDHHS/NIH, 2004: 27). The plan includes short-term goals related to children's environments such as to 'assess children's environments . . . to determine barriers to increasing physical activity' (28); to 'Identify . . . environmental and behavioural factors to obesity . . . prevention . . . and assess . . . environments such as . . . childcare . . . for specific barriers to increasing physical activity' (29).

The severity of the problem has also encouraged regional initiatives in Europe, where approximately 14 million children are already overweight or obese. On 15 March 2005 the European Commission launched the EU Platform for Action on Diet, Physical Activity and Health (EU 2005). A broad spectrum of government, industry and community representatives are part of the initiative (consumer organizations, food industry and health NGOs, among others). Special emphasis is given to programmes for children since this age group shows the highest obesity rise in the region and it is proven that overweight children will become overweight or obese adults.

Targeting and improving children's outdoor environments to support greater amounts of physical activity might be a substantial contribution to the success of these plans. Whole body movement not only influences physical health but general child development, since movement also stimulates brain development (Hannaford, 1995). Moreover, research has confirmed that contact with (and even views of) green environments support attention functioning (Faber Taylor et al., 2001).

It is known that the layout of the site, the number of play settings and the amount and type of vegetation affect children's behaviour (Fjørtoft, 2001; Moore, 1974; Moore and Wong, 1997). The richer the environment, the more engaging for children's play (Grahn et al., 1997). Systematic observations of children's

EVIDENCE-BASED DESIGN FOR YOUNG CHILDREN

9.1 Preschool area tricycle path. BHFS Child Development Center. Research Triangle Park, NC, USA.

interactions with diverse environments confirm these findings. For example, a curvy, 1.5m (5ft) wide pathway can afford children's movement around a circuit of diverse play settings (flower and vegetable gardens, vegetated arbours, circle of rocks, sand play area and others), providing adequate space for continuous roaming and the use of tricycles and carts (Figure 9.1). The level of activity of a play area is mostly due to the additive effect of the layout of the site and its attributes (objects, plants, other children and events) on children's behaviours (Cosco, 2006).

When engaged in self-guided exploration, children stay physically active, performing novel movements and challenging their own developing skills. For example, a group of preschoolers was observed trying to reach a Hyacinth Bean pod (the '*violet thing*') that was hanging high up on a bean tepee (vine-covered metal armature in the shape of a tepee). In this 20-minute episode of activity, six 3-to-5-year-old children (four boys and two girls) were actively engaged without adult intervention in trying to jump up and harvest the intriguing, purple pod. The environment was sufficiently rich and stimulating to support their extended explorations (Figure 9.2).

9.2 Bean teepee planted with a mix of Hyacinth Beans and gourds. Preschool area, BHFS Child Development Center, Research Triangle Park, NC, USA.

The concept of affordance

A key to understanding the implications of the built environment and children's active living is the concept of affordance (Gibson and Pick, 2000). The concept is valuable for describing environments from a behavioural perspective (i.e. from the point of view of children's outdoor play). In this manner, an object in the play area will be considered *climb-able* if it

NILDA G. COSCO

is possible to climb on it, *slide-able* if it allows sliding, or *swing-able* if one can swing on it. The approach considers the individual and the environment as an interactive system.

Children learn about the environment and themselves by picking up environmental information and by performing developmental activities such as climbing, balancing, catching, clinging, crawling, hanging, hopping, jumping, leapfrogging, rocking, rolling, running, skipping, sliding, spinning, walking and so on. However, the environment must be designed to afford these activities. Over time, the daily use of environmental affordances guides future behavioural responses and, as children develop, they learn about the growing scale of their bodies and their emerging specialized skills by using the potential environmental affordances that appear in front of them. The progressive learning and realization of affordances is supported by further environmental exploration that results in sustained activity as perception and action become intimately connected.

The concept of affordance can be utilized, therefore, for discovering and analysing the characteristics of behaviour settings from a young child's point of view (Heft, 1988, 2001). For instance, evergreen plants and grasses automatically add 'pickable' affordances to the environment throughout the year that support rich sequences of play. In an attempt to examine how these interactions work, two preschoolers (a boy and a girl) were observed on a late autumn afternoon carefully picking leaves and collecting them on top of a pail full of sand. They moved in and out of tall grasses around the periphery of the play area, harvesting the 'reachable' leaves and running back and forth to the sand area. They were 'cooking' a birthday party cake. Other girls joined the group activity. When the 'cake' was ready, all walked in a procession-like manner to a picnic table

9.3 Collecting leaves for a 'birthday cake'. The Enrichment Center, BHFS. Research Triangle Park, NC, USA.

9.4 One more leaf for the 'birthday cake'. The Enrichment Center, BHFS. Research Triangle Park, NC, USA.

for the 'birthday party'. This active, cooperative and harmonious group activity would not have occurred without the combined affordances of elements such as pickable grass leaves, pails, sand, picnic table and the story line created by children's imagination (Figures 9.3, 9.4 and 9.5).

Environments full of novel information and rich affordances should be considered as a developmental need to accompany children's growth and the extension of physical capacities (Gibson and Pick, 2000). However, what children perceive is not the abstraction of colour, sound or texture but the *layout* of the space, the *objects* in the layout and the *events* that occur in that particular layout in relation to the existing objects (Gibson and Pick, 2000).

The *layout* contains the surfaces to walk on, the walls or plants that surround subjects, the overhangs that wrap them up spatially and communicate a sense that the body is a volume. The layout of the site helps children to situate themselves in the place that contains *objects* (animate and inanimate) such as people, animals, plants and elements to climb on, sit on, swing on and so on.

According to Gibson and Pick (2000: 24) *events* are 'the movement and actions that occur, some performed by ourselves and some external to us. They implicate objects and provide the dynamics of all scenes in the layout'.

Children learn about their surroundings by performing movements and actions (events), they learn how to orientate themselves using fixed elements such as landmarks, and can increase their territorial exploration with the confidence that they will not be lost. The process involves children's active engagement and supports the emergence of new actions that contribute to expanding environmental experiences. For instance, preschool children are fascinated by wheeled toys. They start by learning to use tricycles, coordinating the movement of their legs, pressing hard on the pedals, and aiming at their destination guided by their arms and hands. Not long after the process starts, they master the movements and can perform other tasks as they drive their wheeled toys. At this stage, they not only ride tricycles but also carry other children with them, along with toys and play materials.

Playing in diverse environments potentially establishes active behaviours in young children and fondness for the outdoors as a preventative measure against sedentary lifestyles in later years.

Physical activity play

Outdoor play is associated with physical activity (Sallis et al., 1993) and higher energy expenditure rates (Pellegrini et al., 1998). Recent articles and governmental websites show that free play has been re-discovered as a critical activity that provides the necessary amount of daily exercise for young children (Dowda et al., 2004; USDA, Nutrition Newsletter). Negating playtime for young children may bring serious health and developmental consequences (Dowda et al., 2004), although children will spontaneously compensate for the lack of play activity when social and physical environments allow for it (Pellegrini and Smith, 1998).

Recently, the report of a panel of experts representing the fields of public health, epidemiology, exercise science, behaviour and medicine was released, containing a review of the current knowledge about physical activity and proposed priorities for research in early childhood (Fulton et al., 2001). The panel acknowledged the importance of play as the main source of physical activity in children two to five years of age, characterized by short bursts of energy such as rough-and-tumble play

9.5 'Birthday cake' on picnic table. The Enrichment Center, BHFS. Research Triangle Park, NC, USA.

(Pellegrini et al., 1998) and group games. The panel highlighted the need for developing reliable physical activity measures for young children and for identifying potential environmental factors including quality of day care (Fulton et al., 2001).

Conditions of outdoor play environments in childcare centres

As a contribution to the baseline knowledge of outdoor environments for young children, the report *Childhood Outdoors:* *A Baseline Survey of Environmental Conditions of Outdoor Areas in North Carolina Childcare Centers* (Cosco and Moore, in press) was recently completed. Approximately 10% of licensed childcare centres in the State of North Carolina, USA were surveyed (n=326). Results show that the large majority of licensed childcare centres offer minimum accommodations for active play beyond basic sand play areas and climbing structures.

Limited environmental diversity reflected in the number of natural or manufactured elements present describes the condition of most outdoor play areas in North Carolina (a

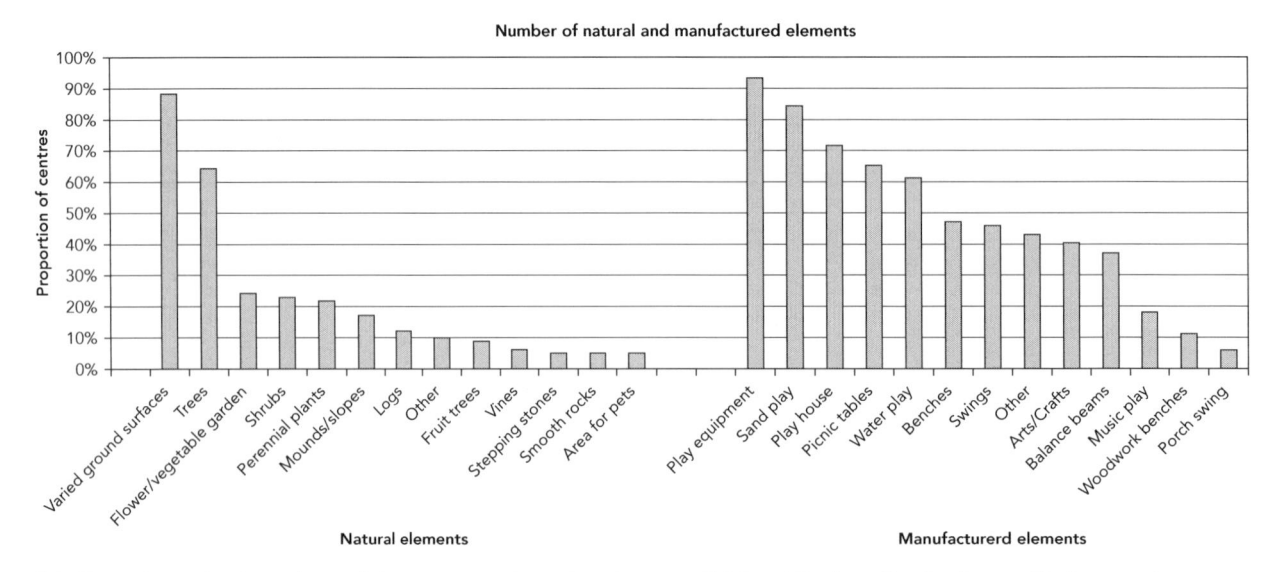

9.6 Types of natural and manufactured elements present in outdoor play areas. *Childhood Outdoors: Baseline Survey of Environmental Conditions of Outdoor Areas in NC Childcare Centers* (Cosco and Moore, in press).

EVIDENCE-BASED DESIGN FOR YOUNG CHILDREN

state considered a leader in childcare standards in the United States). A high percentage of centres provide a single piece of play equipment and few natural elements (Figure 9.6). Shade is seldom provided, even though there is a great concern about the negative health consequences of exposure to the sun. Trees, pergolas and vine-covered arbours are inexpensive shade elements that could be provided but are rarely present.

While the majority of centres have an average of seven manufactured play elements (play equipment, sandbox, play house, picnic tables, water play, benches, swings, easels) the average number of natural elements present is just three (mainly grass, mulch and occasional trees) (Figure 9.7). Lack of diversity is a major reason why outdoor play areas are not attractive.

They are boring and uncomfortable for children as well as for teachers.

The desire for improvement is strong among childcare providers. Respondents to the survey emphasized the need for training and professional help to enhance their outdoor play areas.

Balancing play value and safety regulations

Survey respondents also showed concern about the impact of new health and safety regulations on the quality of the children's outdoor experience. There is a perception among

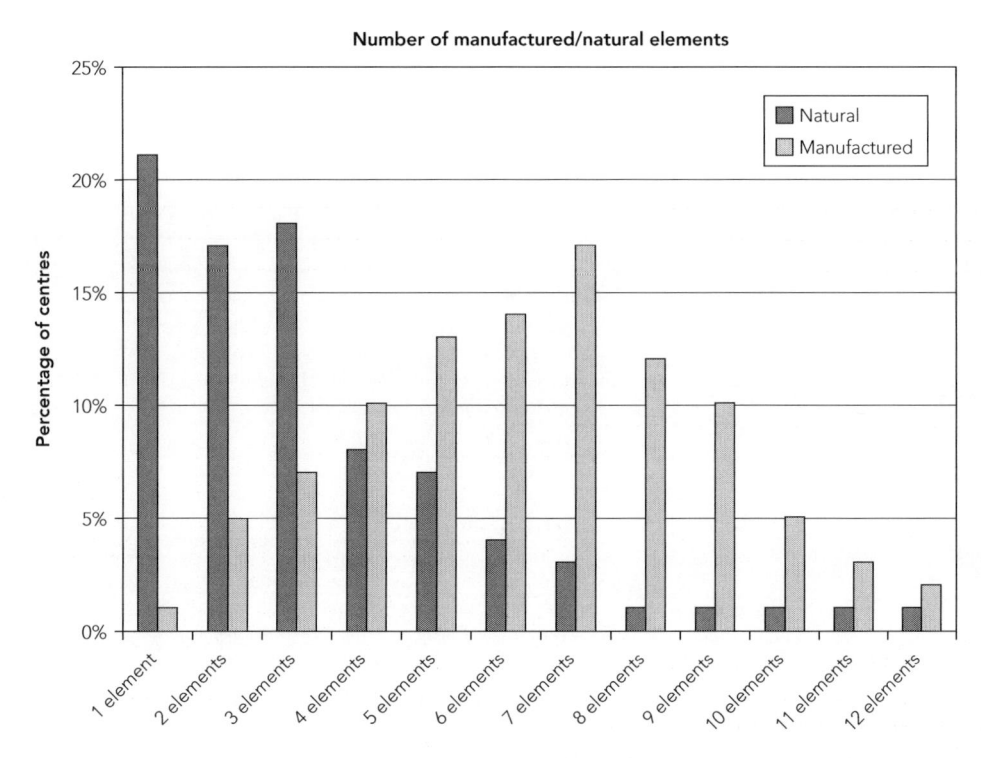

9.7 Number of manufactured and natural elements present in outdoor play areas. *Childhood Outdoors: Baseline Survey of Environmental Conditions of Outdoor Areas in NC Childcare Centers* (Cosco and Moore, in press).

educators and designers that playgrounds in general, and childcare play areas in particular, have turned into unchallenging, un-engaging spaces in the last 20 years. Such environments do not support children's daily requirements for physical activity, and therefore, healthy development. The US Consumer Product Safety Commission CPSC guidelines (*CPSC Guidelines*, 1997) are a significant factor influencing the characteristics of childcare play areas and, unintentionally, have adversely affected their play value. Outdoor play should provide children with the necessary, reasonable risk-taking opportunities that support healthy growth and learning (Frost et al., 2001). Environments that support children's free explorations, expansive movements, and interactions with other children and adults, offer higher play value. The inclusion of childcare facilities within the scope of the *CPSC Guidelines*, attached to the notion of places of 'public use', was presumably intended to address issues of quality of provision, but is a questionable idea. 'Public use' implies a place 'accessible to all members of the community' (Webster's International Dictionary). This is not the case with childcare centres, where children are enrolled in programmes and access is highly controlled. Moreover, during operating hours, close supervision, indoors and outdoors, is provided by trained, professional staff.

Additionally, the way in which the *CPSC Guidelines* are sometimes interpreted and imposed on childcare facilities in the United States has had a clear impact on the layout of childcare outdoor play areas and their patterns of use. For example, the indispensable safety surface under play structures (usually mulch or sand) is often expanded unnecessarily to adjacent areas, transforming them into giant sandpits or mulched areas where it is difficult to run and where wheelchairs or wheeled toys cannot be used. A further issue is the way safety regulations are often enforced by local inspectors without balancing playground safety with the need for high play value.

Research shows that the majority of injuries at childcare centres are minor (no intervention of a doctor is required) and that, although a high percentage occur at the playground (74%), most of them are precipitated by child-related factors (59%) such as pushing and biting (Alkon et al., 1999). Similar results were found by the author in the review of incident reports (minor injuries) conducted before and after renovation of a childcare centre play yard (Cosco and Moore, unpublished). The results showed that after the renovation – that dramatically increased the natural diversity – the number of injuries had a statistically significant decline ($p=0.5$).

However, additional research is needed to confirm these preliminary findings and identify specific physical and social factors that contribute to children's injuries in childcare centres. Lack of empirical research on where and how children get hurt is a major barrier to promoting change in safety regulations for play areas at childcare centres.

To create environments that support children's healthy development and which also comply with licensing and safety regulations requires a sensible understanding and enforcing of the rules. How to confront these two apparently contradictory needs is a current dilemma and suggests the need for policy challenge (Pate et al., 2004). The opinions of educators, parents, owners, safety consultants and designers are divided. Some have chosen to comply with all regulations in the name of safety and have created static, 'equipment-based' play areas surrounded by a sea of wood chips or sand. Others have decided to bypass the safety guidelines (which apply only to anchored equipment) and have created 'garden-like' play areas with minimum or no equipment. Still others have decided to leave selected pieces of equipment and to add trees, shrubs and naturalistic play settings. Surprisingly, the decision-making process is often driven by external constraints (budget, licensing and space limitations) rather than educational or children's health requirements. In any case, there are insufficient research studies to support more rational choices about the quality of children's outdoor environments and their implications for healthy development. This lack of evidence produces a narrow discourse on the topics of safety, environmental diversity and play value.

Need for research

Current research supports the assumption that the childcare centre is an emerging opportunity for successful environmental interventions to counteract the sedentary lifestyles of young children since the childcare centre is the highest predictor of physical activity of children three to five years old (Finn et al., 2002). The fact that preschool physical activity tracks throughout childhood and has a protective effect against early adolescence adiposity (Moore et al., 2003), confirms the hypothesis that the preschool years offer the best opportunity to establish active lifestyles.

More specifically, studies have established that being outdoors is the strongest correlate of physical activity of preschool children (Baranowski et al., 2000; Sallis et al., 1993) and that diverse natural environments support attentional functioning, gross motor development, children's health and richer play behaviour (Sääkslahti et al., 2004; Faber Taylor et al., 2001; Grahn et al., 1997). As a result, researchers now have the opportunity of embarking on the development of studies to uncover the associations between children's physical activity and specific spatial or design attributes with the objective of supporting designers and educators to promote change.

Although the focus of attention to date has been on children regardless of their BMI (Body Mass Index), special consideration should be given to those already overweight (Moore et al., 1995). They might be at increased risk of further weight gains because of low levels of physical activity during the preschool day (Trost et al., 2003) and because they are less inclined to test their physical abilities. A key research topic to address this need (on both sides of the Atlantic) is the study of play setting preferences by overweight children and the description of appealing settings and specific features that might afford sustained or greater amounts of physical activity. A further important focus is low income and ethnic minority children (especially African-American and Hispanic in the USA) and girls, who are more likely to be at risk of being overweight or obese (Mei et al., 1998). Studies that address the use of play areas by these specific populations are urgently needed.

Opportunities for change

Provision of active living environments for young children appears obvious, but there is a need to build a knowledge base through environmental design research to guide policy makers, licensing agencies, designers and teachers. Evidence-based licensing requirements will help emphasize the need to spend time outdoors and reinforce the importance of creating environments that are diverse enough to motivate children and teachers to use them for longer periods of time every day. Empirical evidence could also help lobbying efforts to increase budget allocations for developing outdoor settings to promote physical activity.

The search for evidence-based, site-specific recommendations is currently pursued by governmental health organizations in an effort to counteract the sedentary lifestyle trend and to support the work of planners and designers (DHHS). The places where children spend time daily are a highlighted priority.

In sum, studies that bring knowledge concerning the dynamics of active children's environments will support the creation of new standards of practice. The evaluation of outdoor play areas from the perspective of children's daily physical activity will follow as a natural spin-off and necessary complement to the new standards. For this reason, specific instruments should be developed to measure preschool activity and play area characteristics based on objective research findings.

The *Preschool Outdoor Environments Measurement Scale* – POEMS (DeBord et al., 2005) is an example of a scale intended to measure the overall outdoor quality of preschool play areas and could provide an impetus to develop instruments focused on active lifestyles outdoors. There is no doubt that childcare centres are potential agents of change that could be activated by designing spaces and programmes that support healthy development. Appropriate space design and childcare licensing policies and accreditation regulations can become viable instruments to produce environmental change and, therefore, support healthy behavioural changes in the daily lives of millions of children.

Acknowledgements

This chapter is based on the Doctoral Thesis of the author (*Motivation to Move: Physical Activity Affordances in Preschool Play Areas*, 2006) partially funded by The Robert Wood Johnson Foundation (Active Living Research Program), The Eden Project, United Kingdom, and Bright Horizons Family Solutions, USA.

References

Alkon, A., Genevro, J.L., Tscham, J.N., Kaiser, P., Ragland, D.R. and Boyce, W.T. (1999) 'The epidemiology of injuries in 4 child care centres', in *Arch Pediatr Adolesc Med.*, vol. 153, pp. 1248–1254.

Baranowski, T., Mendlein, J., Resnicow, K., Frank, E. and Weber Cullen, K. (2000) 'Physical Activity and Nutrition in Children and Youth: An Overview of Obesity Prevention', in *Preventive Medicine*, vol. 31, S1–S10.

Capizzano., J., Adams, G. and Sonenstein, F. (2000) *Child Care Arrangements for Children Under Five: Variation Across States* (Research Report). Washington, DC: The Urban Institute.

Cosco, N. (2006) *Motivation to Move: Physical Activity Affordances in Preschool Play Areas.* Unpublished Doctoral Thesis. Edinburgh: College of Art, School of Landscape Architecture. ECA/Heriot-Watt University.

Cosco, N. and Moore, R. (in press) *Baseline Survey of Environmental Conditions of Outdoor Play Areas in North Carolina Childcare Centres.*

Cosco, N. and Moore, R. (unpublished report) *Review of Incident Reports Before and After a Play Area Renovation.* The Natural Learning Initiative, College of Design, NC State University.

Davison, K. and Birch L. (2002) 'Childhood Overweight: A Contextual Model and Recommendations for Future Research', in *Obesity Reviews*, vol. 2, pp. 169–171.

DHHS. *Comparison of Requirements for Child Care Centers with Minimum and Higher Voluntary Program Standards.* Available at www.ncchildcaresearch.dhhs.state.nc.us/reqcomp.htm (accessed 3 January 2006).

DeBord, K., Hestenes, L., Moore, R., Cosco, N. and McGinnis, J. (2005) *Preschool Outdoor Environment Measurement Scale-POEMS.* Winston-Salem, NC: Kaplan.

Dowda, M., Pate, R., Trost, S., Almeida, M.J. and Sirard, J. (2004) 'Influences of Preschool Policies and Practices on Children's Physical Activity', in *J Community Health*, vol. 29 (3), pp. 183–196.

EU (2005) *EU Platform for Action on Diet, Physical Activity and Health.* Available at www.ec.europa.eu/health/ph_determinants/life_style/nutrition/platform/platform_en.htm (accessed 3 January 2006).

Faber Taylor, A., Kuo, F. and Sullivan, W. (2001) 'Coping with ADD: The Surprising Connection to Green Play Settings', in *Environment and Behavior*, vol. 33 (1), p. 54.

Finn, K., Johannsen, N. and Specker, B. (2002) 'Factors Associated with Physical Activity in Preschool Children', in *J Pediatr*, vol. 140, pp. 81–85.

Fjørtoft, I. (2001) 'The Natural Environment as a Playground for Children: The Impact of Outdoor Play Activities in Pre-primary School Children', in *Early Childhood Education J*, vol. 29 (2), pp. 111–117.

Frost, J., Wortham, S. and Reifel, S. (2001) *Play and Child Development.* Upper Saddle River, NJ: Merill Prentice Hall.

Fulton, J., Burgeson, C., Perry, G. and Sherry, B. (2001) 'Assessment of Physical Activity and Sedentary Behavior in Preschool-age Children: Priorities for Research', in *Pediatr Exercise Res*, vol. 13, pp. 113–126.

Gibson, E. and Pick, A. (2000) *An Ecological Approach to Perceptual Learning and Development.* New York: Oxford University Press.

Grahn, P., Mårtensson, F., Lindblad, B., Nilsson, P. and Ekman, A. (1997) 'Out in the Preschool' (Ute på Dagis). *Stad and Land*: 145.

Hannaford, C. (1995) *Smart Moves: Why learning is not all in your head.* Arlington, VA: Great Ocean Publishers.

Health Survey for England (2002). Available at www.dh.gov.uk/PublicationsAndStatistics/.

Heft, H. (1988) 'Affordances of Children's Environments: A Functional Approach to Environmental Description', in *Children's Environments Quarterly*, vol. 5, pp. 29–37.

Heft, H. (2001) *Ecological Psychology in Context.* Mahwah, NJ: Lawrence Erlbaum.

IOTF (2005) *Obesity in Europe Childhood Section. Appendix 1.* Available from www.iotf.org/childhood/euappendix.htm.

Mei, Z., Scanlon, K.S., Grummer-Strawn, L.M., Freedman, D.S., Yip, R. and Trowbridge, F.L. (1998) 'Increasing Prevalence of Overweight Among US Low-income Preschool Children: The Centres for Disease Control and Prevention Pediatric Nutrition Surveillance, 1983 to 1995', in *Pediatrics*, vol. 101 (1), pp. 103–105.

Moore, R. (1974) 'Patterns of Activity in Time and Space: The Ecology of a Neighborhood Playground', in D. Cantor and T. Lee (eds) *Psychology and the Built Environment.* London: Architectural Press, pp. 118–131.

Moore, R. and Wong, H. (1997) *Natural Learning: The Life History of an Environmental Schoolyard.* MIG Communications.

Moore, L., Gao, D., Bradlee M. and Cupples, L. (2003) 'Does Early Physical Activity Predict Body Fat Change Throughout Childhood?', in *Prev Med*, vol. 37 (1), pp. 10–17.

Moore, L., Nguyen, U.D.T., Rothman, K.J., Cupples, L.A. and Ellison, R.C. (1995) 'Preschool Physical Activity Level and Change in Body Fatness in Young Children. The Framingham Children's Study', in *Am J Epidemiol*, vol. 142 (9), pp. 982–988.

Ogden, C., Flegal, K.M., Carroll, M.D. and Johnson, C.L. (2002) 'Prevalence and Trends in Overweight among US Children and Adolescents', in *JAMA*, vol. 288 (14), pp. 1728–1732.

Pate, R., Pfeiffer, K.A., Trost, S.G., Ziegler, P. and Dowda, M. (2004) 'Physical Activity among Children Attending Preschools', in *Pediatrics*, vol. 114 (5), November 2004.

Pellegrini, A. and Smith, P. (1998) 'Physical Activity Play: The Nature and

Function of a Neglected Aspect of Play', in *Child Development*, vol. 69 (3), pp. 577–598.

Pellegrini, A., Horvat, M. and Huberty, P. (1998) 'The Relative Cost of Children's Physical Play', in *Animal Behavior*, vol. 55, pp. 1053–1061.

Piaget, J. (1952) *The Origins of Intelligence in Children*. London: Routledge & Kegan Paul.

Reilly, J., Jackson, D.M., Montgomery, C., Kelly, L.A., Slater, C., Grant, S. (2004) 'Total Energy Expenditure and Physical Activity in Young Scottish Children: Mixed Longitudinal Study', in *The Lancet*, vol. 363, pp. 211–212.

Sääkslahti, A., Numminen, P., Varstala, V., Helenius, H., Tammi, A., Viikari, L. and Valimaki, I. (2004) 'Physical Activity as a Preventive Measure for Coronary Disease Risk Factor in Early Childhood', in *Scandinavian Journal of Medicine Science Sports*, vol. 14 (3), pp. 143–149.

Sallis, J., Nader, P.R., Broyles, S.L., Berry, C.C., Elder, J.P. (1993) 'Correlates of Physical Activity at Home in Mexican-American and Anglo-American Preschool Children', in *Health Psychology*, vol.12, pp. 390–398.

Sallis, J., Prochaska, J. and Taylor, W. (2000) 'A Review of Correlates of Physical Activity of Children and Adolescents', in *Medicine Science of Sports Exercise*, vol. 32 (5), pp. 963–975.

Sonenstein, F., Gates, G., Schmidt, S. and Bolshun, N. (2002) 'Primary Child Care Arrangements of Employed Parents: Findings from the 1999 *National Survey of America's Families*' (Vol. Occasional Paper). Washington, DC: The Urban Institute.

Trost, S., Sirard, J., Dowda, M., Pfeiffer, K. and Pate, R. (2003) 'Physical Activity in Overweight and Nonoverweight Preschool Children', in *Int J Obesity*, vol. 27, pp. 834–839.

US Consumer Product Safety Commission (1997) *Public Playground Safety Handbook*. US Consumer Product Safety Commission.

US Department of Health and Human Services, National Institutes of Health (USDHHS/NIH) (2004) *Strategic Plan for NIH Obesity Research. A Report of the NIH Obesity Research Task Force*. NIH Publication No. 04–5493.

US Department of Agriculture, Food and Nutrition Service. *Nutrition Newsletters for Parents of Young Children*. Available at www.fns.usda.gov/tn/Resources/Nibbles/childs_play.pdf.

US Surgeon General (1996) *Report on Physical Activity and Health At-a-Glance Summary*. Available at www.cdc.gov/nccdphp/sgr/ataglan.htm.

Webster's Third New International Dictionary (1981) US: Merriam-Webster.

's
Healing gardens for people living with Alzheimer's

Challenges to creating an evidence base for treatment outcomes

John Zeisel

Introduction

Common sense and clinical experience tell us that walking and being otherwise active in a safe and well-planned garden is therapeutic for people living with Alzheimer's. They get physical exercise, are in a better mood, have a better sense of time and seem to sleep better at night. The entire experience of being outside with nature affects people in this way, not only the daylight. However, knowing this and proving it are two different questions. This chapter explores the challenges that face researchers who want to develop an evidence-base to prove that a well-designed healing garden represents a clear and demonstrable treatment for Alzheimer's disease.

Background

Alzheimer's is a degenerative disease with a quite specific aetiology. A protein called *plaque* covers parts of the brain while dendrites of individual cells disintegrate into what are called *tangles*.

Alzheimer's plaques and tangles over time affect similar parts of the brain of people living with this disease, although at different times in its progress. After 10–15 years living with the disease, up to 40% of a person's brain weight, and as much of its cellular structure, can be affected. At the same time, during the years living with the disease, as much as 90% or 80% is still functioning. The good news is that these working parts of the brain provide us with the key to the built environment's 'treatment effects', including health effects of contact with the outdoors through gardens and their designs.

Among the parts of the brain that hold the keys to healing garden treatment are the hippocampus, the amygdala, executive function in the frontal lobe and the suprachiasmatic nucleus. The following simplified description illustrates the logic and process that successful healing garden design follows.

The *hippocampus* is a small, sea-horse-shaped organ located next to the almond-shaped amygdala in the 'limbic' area of the brain. In Greek, seahorse is hippocampus and almond is amygdale. Sometimes called the key to the brain's files, one function of the hippocampus is to tag each experience and distribute it into the brain's memory bank so that people can retrieve memories when they need them – called recall. This organ in people living with Alzheimer's is often damaged early in the disease, making it difficult to retrieve old memories and to tag new ones. Of particular significance to garden design is the fact that memories of place and location, held as 'cognitive maps', are among the memories that are hard to tag, embed and retrieve. Unable to 'remember' what is around a corner or behind a hedge, a person living with Alzheimer's may find a complex garden or other outdoor space confusing and frightening. A garden designed to help a person find their way without using cognitive mapping capacities that they have lost, is a successful healing garden. Successful design for this group of users can be accomplished by employing natural mapping techniques (Norman, 1988) and hard-wired landmarks.

The brain's *amygdala* handles emotion, feelings and moods and sits right next to the hippocampus. This organ is damaged quite late into the disease, enabling people living with Alzheimer's to remain exquisitely sensitive to emotional subtleties and expressions for a long time. A garden that elicits positive moods that people living with Alzheimer's can fully experience and by which they can orientate themselves, is a successful healing garden. Positive moods, including feeling safe, pleasure and competence, can be achieved by using landmark orientating elements, clear pathways and enclosure that prohibit access to dangerous areas and views that might create frustration.

Frontal lobe damage to 'executive function' centres in the brain leads people living with Alzheimer's to have difficulty organizing complex – or even not so complex – sequences of events. Indoors, such sequenced events might be cooking a meal, brushing teeth or cleaning up. In relation to the outdoors, a person needs executive function to find the way outside, plan a walk and organize a garden activity. A self-organizing outdoor environment enables a person living with Alzheimer's to use a garden without having to mentally organize him- or herself. A successful healing garden presents necessary sequences in obvious and self-evident visual and experiential ways.

The suprachiasmatic nucleus (SCN) is generally considered the body's master circadian 'clock'. Outdoor and indoor light signals adjust the SCN, which then relays this time-information to the rest of the body (Wikipedia Dictionary). While the inner clock that the SCN helps the body maintain is not exactly a 24-hour clock, it helps us sleep at night, wake in the morning and know the difference between night and day, even if we have little contact with the sun and the weather. As a result of damage to their time-keeping ability, people living with Alzheimer's without cues as to time of day, day of week or season have time-related problems. Cues that include frequent contact with external weather and other conditions in a garden can contribute to reduced 'sleep/wake disturbances' in which a person waking at night does not understand it is night time, and 'sundowning' in which a person living with Alzheimer's may become agitated at the end of the day and say they 'want to go home' even if they are already at home. Experience with people living at the Hearthstone Alzheimer Care assisted living treatment residences and using the gardens described below demonstrates that contact with natural elements such as cold and hot weather, sunshine and snow, through on-going access to a healing garden, can reduce such symptoms.

The following healing gardens are successful in their design, in that they respond to the needs described above.

Garden at Hearthstone Alzheimer Care at New Horizons, Marlboro, Massachusetts

The Hearthstone at New Horizons garden is a large, triangular shaped garden on the ground floor of a three-story Assisted Living residence adjacent to the living and dining rooms, on a grade that people living with Alzheimer's use during the day (see Figure 10.1). Because a 2.4 m (8 ft) fence surrounds the garden, providing complete security for residents, staff members do not worry about safety. Residents therefore have essentially 24-hour access to the healing garden from the living room, through a residential 'front door' with glass windowpanes and sidelights that enable residents inside to see what is taking place in the garden outside. Of course, in really bad weather, the door is locked.

Once outside, residents see and experience several distinct areas: a front porch to the right, a back patio to the left and a clearly marked pathway that starts at the right and circles the garden until it reaches the other side of the back patio.

HEALING GARDENS FOR PEOPLE WITH ALZHEIMER'S

Garden and Common Spaces at Hearthstone at New Horizons, Marlboro, MA, USA.

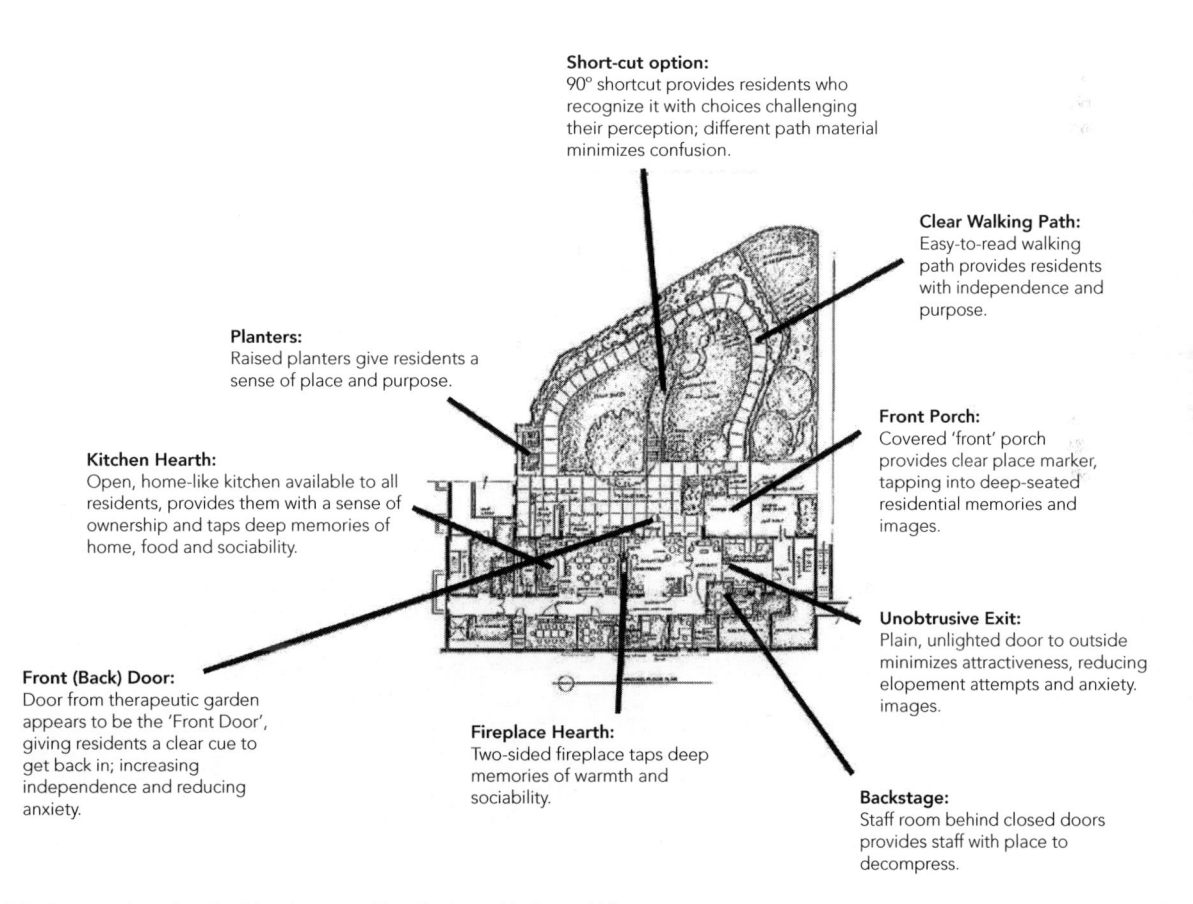

Short-cut option:
90° shortcut provides residents who recognize it with choices challenging their perception; different path material minimizes confusion.

Clear Walking Path:
Easy-to-read walking path provides residents with independence and purpose.

Planters:
Raised planters give residents a sense of place and purpose.

Front Porch:
Covered 'front' porch provides clear place marker, tapping into deep-seated residential memories and images.

Kitchen Hearth:
Open, home-like kitchen available to all residents, provides them with a sense of ownership and taps deep memories of home, food and sociability.

Unobtrusive Exit:
Plain, unlighted door to outside minimizes attractiveness, reducing elopement attempts and anxiety. images.

Front (Back) Door:
Door from therapeutic garden appears to be the 'Front Door', giving residents a clear cue to get back in; increasing independence and reducing anxiety.

Fireplace Hearth:
Two-sided fireplace taps deep memories of warmth and sociability.

Backstage:
Staff room behind closed doors provides staff with place to decompress.

10.1 Annotated garden plan (Hearthstone at New Horizons, Marlboro, MA).

The path circles a 'front yard' and a 'back yard' and passes small garden areas, a 'park bench area' off the path, and two raised planting beds set in another surface material (see Figures 10.2–10.7). The 'front door' – actually the garden door leading back into the living room of the residence – is set into a peak-roofed enclosure that clearly indicates 'way in' to residents using the garden (see Figure 10.6). The garden is surrounded on two sides by three-storey high buildings, and on the third side by the 2.4 m high fence, the top 0.6 m (2 ft) of which is decorative.

10.2 View from 'the park': pathway, short cut, landmark and 'front door' (Hearthstone at New Horizons, Marlboro, MA)

10.3 Bird feeder as active mini-landmark (Hearthstone at New Horizons, Marlboro, MA).

10.4 Rural mailbox (occasionally with mail) and seating area next to 'the park' as interactive mini-landmark (Hearthstone at New Horizons, Marlboro, MA).

10.5 Planting at intersection of short cut and main path creates orienting 'node' (Hearthstone at New Horizons, Marlboro, MA).

10.6 Iconic 'front door' and landmark arbour are readily recognized as a re-entry to home, increasing independent use of the garden (Hearthstone at New Horizons, Marlboro, MA).

10.7 The 'park' with benches backed to fence provides an orienting destination as well as creating a feeling of safe enclosure (Hearthstone at New Horizons, Marlboro, MA).

Garden at Hearthstone Alzheimer Care, the Esplanade, White Plains, NY

This 6 by 24 m (20 by 80 ft) roof-top garden and the adjacent 29-person Hearthstone Alzheimer's Care Assisted Living residence take up the entire floor of a 15-storey building in White Plains, New York, just north of New York City. The garden is long and thin, with one entrance from the kitchen/dining room and another at the other end, through an activity room (see Figure 10.8). Enclosed by a 1.8–2.4 m (6–8 ft) scalloped, light blue anodized metal fence made of 50 by 12 mm (2 by 0.5 inch) slats, with 25 mm (1 inch) of viewing space between each slat, the garden is totally secure.

As with the Massachusetts garden, distinct areas are planned so that residents feel different in each: a covered front porch outside the kitchen, a back yard with seating and a barbecue outside the second doorway, and a 'park' with half-round benches between these two areas. A curved pathway painted onto the pavement connects all three areas. Physical elements define each of these areas: a covered arbour, picnic tables, garden chairs and benches (see Figures 10.9 and 10.10).

The intermediate 'park' is separated from the other two areas by arched rose arbours (see Figure 10.11), and on both sides of the garden pathway is a different coloured pavement, indicating where large potted shrubs and trees are to be put (see Figure 10.12).

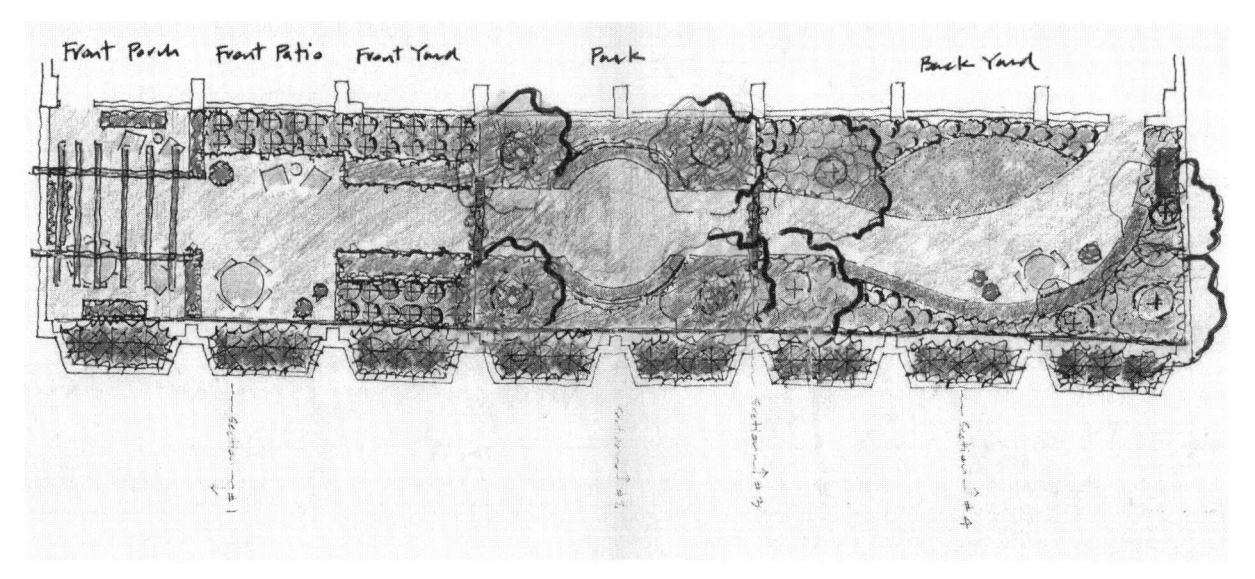

10.8 Conceptual sketch: three-part garden provides wayfinding clarity (Hearthstone Alzheimer Care at the Esplanade, White Plains, NY).
Credit: Garden plan, Martha Tyson.

10.9 Iconic arbour provides clear image to the front porch (Hearthstone Alzheimer Care at the Esplanade, White Plains, NY).
Credit: Sketch, Martha Tyson.

Design principles

Three overlapping design schema underlie successful healing garden design for people living with Alzheimer's (Zeisel and Tyson, 1999). These are:

- natural mapping (developed by Donald Norman (1988) in *The Psychology of Everyday Things*);
- latent image elements (developed by Kevin Lynch (1960) in *The Image of the City*);
- universal housing zones (developed by John Zeisel and Polly Welch (1981) in *Housing Designed for Families*).

Naturally mapped environments and objects have all the information needed to understand their organization and use them embedded into the object or environment itself. No instruction book, map or memory is needed to negotiate the environment or figure out how to make the object work (Norman, 1988). A naturally mapped garden is one which has a few clearly recognizable pathways that can be seen from anywhere in the setting, landmarks indicating transitions between environmental events, evident destinations that users of the environment can see easily, and an exit that everyone can see and understand as the place to leave the garden. A garden that is not naturally mapped would have several forks in the pathways leading to destinations that are hidden around curves and bushes, that leave users in places with no clear way out, and might even have paths that lead back on themselves without an indication of a way out.

Latent image elements are physical attributes of settings that correspond to the brain's natural cognitive mapping abilities. Kevin Lynch, in his landmark study and book *The Image of the City*, interviewed taxi cab drivers in Boston – a complex non-linear grid environment – to understand how they mentally organize information about the city in which they work. The five elements Lynch found were:

- *Paths:* the channels along which people move; the predominant element in their image of a garden as they move through it (see Figure 10.2).
- *Edges:* boundaries between two areas; real or symbolic barriers between areas and transitions that join parts of a garden together. Edges such as the fence around a garden define and hold together general areas (see Figure 10.7). Edges along the sides of a path distinguish the path from adjacent districts and signal transitions from one to the other.
- *Districts:* sections of a garden that someone can pass by or walk into. Garden districts such as front yards or back patios are, like a neighbourhood in an urban area, recognizable as having a unique identifying character (see Figure 10.12).
- *Nodes:* places in a garden that are a focus to and from which people travel. The crossing of two paths, junctions of paths and districts, and places of intense activity represent nodes. Nodes indicate shifts in a movement pattern and choices for users (see Figure 10.5).

JOHN ZEISEL

10.10 Park bench with safe fencing provide destination and freedom (Hearthstone Alzheimer Care at the Esplanade, White Plains, NY).

10.12 Garden with flowers (Hearthstone Alzheimer Care at the Esplanade, White Plains, NY).

10.11 Iconic archways provide clear transitions between the 'park' and front and back porches (Hearthstone Alzheimer Care at the Esplanade, White Plains, NY). *Credit: Sketch, Martha Tyson.*

- *Landmarks:* landmarks are reference points that users of a setting employ to orientate themselves (see Figures 10.3 and 10.4). Towers, domes, signs, trees and doorways are all potential landmarks that are 'increasingly relied upon as a journey becomes more and more familiar' (Lynch, 1960: 48).

These elements are central to the way the brain processes environmental place information. Research has shown that landmarks play a critical role in how people and other animals organize mental information for wayfinding – how they develop their cognitive maps.

Universal housing zones make up the naturally evolved sequence of spaces in and around culturally responsive individual houses and collective housing. Such housing is naturally organized in the sequential space zones set out below (see Figure 10.13) (Zeisel and Welch, 1981). Where the indigenous housing spatial typology is fully realized, that residential setting feels most comfortable and safe. The typology reinforces what may well be a 'hard-wired' sense of home we hold in our brains, although further research is needed to test this neuroscience hypothesis. Each zone can either be realized by space allocated to it – such as a front yard – or by another physical element inserted in place of the zone – such as a steep change of grade, a fence or thick planting. When a zone is completely omitted in design, residents experience environmental stress: feelings of invaded privacy, loss of control and lack of security.

- *Outsider public:* the place in a neighbourhood where everyone is welcome and feels comfortable, no matter if

HEALING GARDENS FOR PEOPLE WITH ALZHEIMER'S

they live nearby or are visiting from afar – like a public park. In a garden, such a place can be designed at the corner of the garden furthest from the entrance.
- *Insider public – front:* a place such as a street between two rows of houses where everyone is free to walk but where those who live along the street keep a close eye on strangers who might not belong. Jane Jacobs (1961) identifies such 'eyes on the street' as essential to a healthy community.
- *Personal areas – front* include gardens and lawns that belong to someone and, whether fenced in or not, are physically accessible to passers-by. When children playing in the street hit a ball into an adjacent front lawn, for example, it is acceptable for them to cautiously fetch the ball themselves.
- *Building edges – front* are habitable and usually built places that extend the interior of a house outside, clearly belong to and are used by the residents who live there, and are off-bounds to strangers with no business there. Front building edge elements such as porches and front stoops are physically accessible and visible to others, but only residents, invited guests and delivery people are supposed to be there.
- *Building walls:* the exterior shell of a house with its 'screening' devices of windows and doors, separates the interior of the house from exterior building edge elements. Ground floor openings in the building wall – windows and doors – require different design qualities to windows above grade.

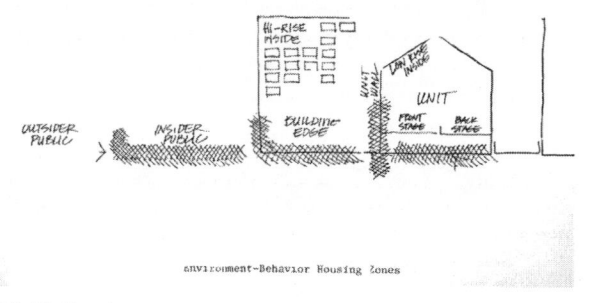

10.13 Housing zones.
Credit: Zeisel and Welch, Housing Designed for Families, 1981.

- *Front stage* interior areas tend to include the more formal welcoming areas of a home such as the living room and parlour (Goffman, 1959).
- *Back stage* areas in the home are more informal rooms in which people relax and prepare to 'go out into the world' – to go 'on stage'. These include an eat-in country kitchen, the family room or den, and bedrooms.
- *Building edges in the back* of the house include patios and back porches that are more private than front building edge elements and therefore clearly off-bounds to outsiders. They extend back stage areas of a home to the outside.
- *Personal areas – back* are used exclusively by residents and their invited friends, such as back yards where children play, family barbecues are held and neighbours gather.
- *Insider public areas in the rear:* a place that residents use for transportation and circulation, as well as for 'back-stage' public tasks such as rubbish collection and equipment storage. Back insider public spaces include back alleys and other short cuts shared mainly by those who live there.

Successful residential plans and Alzheimer's healing gardens include all these zones, either by spatial or some other defining element, such as a change of grade or a fence.

In sum, where all three schemas are designed into a garden – latent image elements naturally mapped in universal (possibly hard-wired) housing zones – that garden has the greatest chance of providing people living with Alzheimer's the most accessible, easily understood and used outdoor spaces. They are truly healing gardens.

Evaluating the effects of healing gardens for Alzheimer's

The last sentence: 'They are truly healing gardens', is more a statement of belief than one with an evidence base. Unfortunately, while there is much clinical, experiential and anecdotal evidence that what I have written in the first part of this chapter is in fact true, there is little rigorous research evidence. Why?

Research has demonstrated that certain interior design characteristics of residences for people living with Alzheimer's are associated with reduced symptoms of the disease. Camouflaged doors that staff and visitors use but that are not intended for residents' independent use, uniquely designed common areas and appropriate residential décor are associated with reductions in social withdrawal, agitation, aggression and delusions among people living with Alzheimer's (Zeisel et al., 2003). Healing or treatment gardens are intended to have similar effects – reduced sleep-wake disturbances, less 'sundowning', better self-regulated internal mental/body clocks of people living with the disease. These results have been observed in clinical situations. Given what we know about the brain, how Alzheimer's affects it and how environments affect the brain, we believe these to be true. Proving these intended environment–behaviour effects with rigorous research represents the next challenge to be overcome.

Defining and describing a particular healing/treatment garden is a challenge. A healing garden is not merely a space. It is a place someone manages. It may be available all day long or only at certain times; it may have organized activities (frequently or seldom) or not; it may have a see-through or an opaque enclosure; it may have a completely secure enclosure, or one that can be easily climbed over. The degree to which the garden is visible from inside can lie anywhere on a complex continuum. The list of variables is long; long enough to pose theoretical, methodological and practical problems for environment–behaviour evaluators.

Theoretical/methodological issues: in a cause–effect evaluation model, researchers identify five types of variable, describe the situation under study in terms of those variables, employ methods they trust to measure the qualities and quantities of each variable and analyse the interrelationships, determining a complex set of causes and effects in the situation. The five variable types are independent, contextual, intervening, dependent and secondary or side-effect.

- *Independent variables* are the characteristics of the situation that will be studied as causes. While they may be the result of some other set of processes (variables) in the situation under study they are seen as the starting point for the evaluation. The characteristics of a garden the effects of which one wants to evaluate – such as size, enclosure, paving, plantings – would be independent variables.
- *Contextual variables* describe the organizational and behavioural context of the environment that must be considered in the evaluation, but are considered external to the study. In an environmental post-occupancy evaluation, whether or not a particular assisted living residence is a for-profit or not-for-profit organization and whether a corporation or an individual owns it would be contextual variables.
- *Intervening variables* are those that indirectly affect the cause–effect relationships being studied, but are not being examined for their direct effects. For example, the weather during the period a garden is being evaluated, or a change in care staff, can have dramatic effects on how the garden is used in an assisted living residence, influencing behaviours and attitudes that constitute the outcomes one is interested in.
- *Dependent variables* are those that the evaluation treats as outcomes affected by changes in values of the independent – causal – variables. The quality and quantity of these variables are 'dependent' on the qualities and quantities of the 'independent' variables. Dependent variables can be mental states, cognitive abilities, moods, perceptions, behaviours and attitudes, among others.
- *Side-effect variables* are those characteristics of users – behaviours, moods and so on – that the independent variables affect indirectly through changes in the values of the dependent variables. For example, the safety designed into a garden (independent variable) may enable staff to leave the doors open for long periods of time (dependent variable), enabling residents to come and go as they please (dependent variable), rather than only with the assistance of a caregiver. The resulting side-effect is a greater feeling of independence among users.

An interesting observational study that was carried out at the Hearthstone at New Horizons treatment garden in Marlboro sheds light on one important side-effect of having a well-designed garden – increased resident independence. Grant (2003) observed residents, family members and staff who used the garden over a period of six days in the summer of 2001. She uncovered two enlightening uses of the garden, both representing contributions to residents' feeling of independence.

First, she found that the most used single element (13%) was the 'park bench' in the 'outsider public' part of the garden. Residents sat on a bench alone or with family. This solitary element and district was the most used single object/place in the garden.

Second, residents on their own initiated 36% of all trips into the garden from inside, not accompanied by anyone else or as part of a group. Fully 59% of the trips back inside were alone. The naturally mapped garden, designed by Martha Tyson to include Lynch's image elements and Zeisel and Welch's universal housing zones (Tyson, 1998), gave residents living with Alzheimer's the opportunity to come and go as they pleased. This unintended effect is a major accomplishment for such an environmental element, and one which future research will surely evaluate.

The large number of variables in each of the above-listed categories, and the complex values each can take, underlies the methodological challenges garden post-occupancy evaluation researchers must overcome if we are going to be able to identify which healing gardens provide true treatment and why.

Developing measures to assess outcomes and side-effects is difficult, for several reasons, as outlined below.

Controlling for spurious and other effects: rigorous cause–effect evaluation research enables analysts to identify which combination of independent variable characteristics are responsible for the observed changes to the dependent variables. To achieve this, contextual and other extraneous variables must somehow be controlled, either by establishing a control group or by statistically controlling these variables in data gathering and analysis. Achieving this is extremely difficult when the number of variables in a given situation is overly complex, and this is the research dilemma gardens present us with.

Measuring specific exposures: even measuring a person's exposure to daylight in the garden can yield little clear data. Light monitors worn on the wrist, for example, also record daylight through windows in the interior, making data analysis difficult. This is only one of the many complexities that make it difficult to assess the actual effects of healing gardens on people living with Alzheimer's. Overcoming these methodological difficulties is a major challenge (and opportunity) for the environmental psychology and landscape design communities.

Independent variables in the garden environment

The first step in addressing the research question: *What are the effects of a healing garden on someone living with Alzheimer's?* – even before identifying the effects one wants to study – is to *describe the characteristics of the healing/treatment garden or gardens* one is interested in studying.

- How big is the garden? Is it like a small outside room or like an adjacent neighbourhood containing several smaller districts and rooms? If it is small – like an exterior room – is it a single undifferentiated space, or are there several smaller areas in the space? Is the garden big enough that someone using the place furthest from the building entry feels he or she is 'away' from the building, or does a user feel he or she is still near the building, with no feeling of getting away in the garden?
- Is the garden merely an outdoor space or is it an outdoor area designed according to the three principles of natural mapping (Norman, 1988), five cognitive mapping elements (Lynch, 1960) and universal housing zones (Zeisel and Welch 1981)? If designed in this way, how many of the mental cueing elements does it contain and how are they arranged and related to one another?
- Are there plantings in the garden: grass, trees, shrubs, flowers and vegetables? Are the trees mature, giving a sense of stability, or newly planted? Do the flowers bloom throughout the flowering season or are they short-blooming?
- What objects are there in the garden to attract residents' attention? Are there benches to sit and observe, raised flower and vegetable planting boxes that residents can tend, bird feeders that can be observed, mailboxes that attract attention, small concrete animals that amuse or even an old car that may encourage memories?
- If there are activity areas – Lynch's districts and Zeisel and Welch's housing zones – are these designed and constructed appropriately for their use? Is the backyard patio paved so that it is easy to have a barbecue? Is the porch situated and furnished so that there is activity to overlook and passively enjoy?

- How is the garden enclosed and how does the enclosure affect the garden's security for residents? Is the surrounding fence high enough – 2.4 m (8 ft) – to entirely prevent residents from climbing out of the garden and ending up in a place they cannot understand, and that is therefore dangerous? Or is the fence low and decorative, providing little safety protection? Is the fence solid – such as wood planks with no space between them – or is it a see-through fence that invites users to join the attractive yet potentially dangerous activities beyond the garden?
- Does the shape of the garden leave all garden areas observable from the inside or could a resident fall and hurt him- or herself unobserved in an unseen corner? (If so, staff are likely to keep the door to the garden locked so that that can't happen.)
- Are the paths wide enough for two people to walk side-by-side with many wayfinding landmarks and no confusing 'forks in the road'?
- What is the relationship of the garden to the interior of the Alzheimer's residence? Through what room do residents enter the garden, and is there anything in the door's design that makes it evident to residents that this is the way out, and from the garden, the way in?

If these were not enough design characteristics to have to identify in specifying the gardens being studied, the task becomes even more daunting when we add the complexity of identifying the intervening factors that influence garden effects on residents.

Intervening factors

The most significant *intervening factors* are the gate-keeping system and planned activity programmes.

At one end of the spectrum are gardens that are always open for use, so that any resident can walk around or sit in the garden whenever she or he wants. At the other end is a garden whose door is always locked except when a staff member opens it and accompanies one or more residents outside. In almost all healing gardens for people living with Alzheimer's, there is a formal or informal gatekeeper who determines when the door is open. The gatekeeper can vary by attitudes towards nature, whether she or he is a staff member, a family member or if access is regulated by a policy that everyone knows to follow, such as 'official opening hours'. Another factor is whether the garden is used during all seasons or just in the milder ones. In northern climates, residents can always shovel snow on the pathway in winter if they are warmly dressed.

The degree to which residents can decide on their own to use the garden or if they can only go to the garden accompanied by a staff or family member, significantly impacts resident outcomes (Grant, 2003).

And then there is the question of activities. Are there many or few planned activities? Are they independent of there being a garden – like playing cards – or are they related to being outdoors, such as cultivating, planting, weeding and harvesting tomatoes, or maintaining the garden by raking leaves and shovelling snow?

User mix also affects outcomes. Is the garden used by everyone living in the adjacent housing, just for elders or just for elders who are living with Alzheimer's?

Dependent variables: treatment outcomes and their measurement

The symptoms of Alzheimer's, primary symptoms such as difficulties finding words to express oneself and organizing complex sequences, as well as secondary and tertiary ones, including social withdrawal and aggression, can all be affected by the presence of a garden and its design. The decision as to what variables to select as dependent variables in a post-occupancy evaluation study of gardens presents a major hurdle.

Outcomes research needs to focus not on the kind of brain-based symptoms described at the beginning of this chapter, but on the behaviours, functional abilities and cognitive skills the person has and that the physical environment they inhabit might influence, including gardens.

Among the behaviours that are the focus of such evaluations, the four Alzheimer's 'A's stand out.

- *Anxiety* – worrying about things that can't be controlled.

- *Agitation* – worrisome physical and verbal actions disruptive to others.
- *Aggression* – striking out at others invoked by perceived aggression.
- *Apathy* – lack of involvement invoked by a boring environment.

Gardens can affect the amount of each of these behaviours a person living with Alzheimer's exhibits, because these are affected by the degree of meaningful activity the environment provides – one of the intervening variables in garden evaluation – and also the degree to which the environment is naturally mapped. Measures of agitation and aggression are included in a valid and reliable research scale, the Cohen-Mansfield Agitation Inventory (Cohen-Mansfield et al., 1989). Apathy is not often included among the symptoms of Alzheimer's, but the work of Ladislav Volicer has brought this symptom to the fore. Volicer points out (Volicer and Bloom-Charette, 1999) that although apathy – sitting around doing nothing – is not disturbing to others, it is a real symptom of environments that provide inadequate meaningful activity for people living with Alzheimer's who have difficulty generating activities themselves.

Among other behaviours that can be selected as dependent outcome variables are the tendency to withdraw socially (measurable with the *Multidimensional Observation Scale for Elderly Subjects* (Helmes et al., 1987)) and mood. Mood can further be elaborated into joy, self-confidence, feelings of independence and reduced depression, the latter also measurable through the *Multidimensional Observation Scale for Elderly Subjects*.

Functional abilities include the ability to walk by oneself, get dressed, take a shower, eat by oneself and even activities such as gardening, raking and weeding. Gardens that explicitly provide garden-related activities can be evaluated in terms of those, and might even have side-effects in the other functional areas.

Physical environments may also affect cognitive skills, although these variables are more tenuously connected to gardens and garden design. They include increased memory recall and learning, longer focus of attention, increases in self-awareness and language ability – the ability to understand and express words – and fewer psychotic problems such as misidentification and paranoid delusions. The latter are measurable with the *Reisberg Psychotic Symptom List* (Reisberg et al., 1987).

Physical health is the last of these independent variables that gardens may well affect. Balance, physical strength, mobility, lack of physical discomfort and perhaps even fewer infections are among possible outcomes of being outside more and participating in garden activities.

Altogether, these variables contribute to residents' quality of life – a global aggregate variable in its own right.

Finally, with innovations in measurement technologies, the neurosciences increasingly come into play when we identify possible brain-based evidence for environmental design improvements. For example, neuroscientists know that each cell in the hippocampus – such as ensemble activity in Area CA1 – is affected by the spaces in which we find ourselves. The future of environmental outcomes research will certainly include neuroscientific methods (Zeisel, 2006).

Summary

While the theory and practice of healing garden design is quite advanced, post-occupancy evaluations of such places still have significant methodological challenges to overcome if they are to demonstrate the treatment effects of healing gardens.

This chapter makes no attempt to resolve these difficulties; rather, it attempts to identify the many research variables and their values that need to be considered so that future researchers can unravel the mysteries of healing/treatment gardens and shed light on their healing/treatment effects.

The chapter does not argue that we ought not to evaluate the positive effects many of us know treatment gardens have; rather, that caution is required before we cavalierly start measuring garden effects without careful methodological development. If we do not engage in this development first, we may find that we cannot identify actual effects adequately and such negative findings eventually return to haunt us.

References

Cohen-Mansfield, J., Marx, M.S. and Rosenthal, A.S. (1989) 'A description of agitation in a nursing home', in *J of Gerontology*, vol. 44, pp. 77–84.

Goffman, Erving (1959) *The Presentation of Self in Everyday Life*. New York: Doubleday.

Grant, Charlotte (2003) 'Chapter III: Hearthstone at New Horizons', in *Factors Influencing the Use of Outdoor Space by Residents with Dementia in Long term Care Facilities*. PhD Thesis, Georgia Institute of Technology, UMI Dissertation Publishing. Available at disspub@umi.com: www.umi.com.

Helmes, E., Csapo, K.G. and Short, J.A. (1987) 'Standardization and validation of the Multidimensional Observation Scale for Elderly Subjects (MOSES)', in *J of Gerontology*, vol. 42, pp. 395–405.

Jacobs, Jane (1961) *The Death and Life of Great American Cities*. New York: Vintage.

Lynch, K. (1960) *The Image of the City*. Cambridge, MA: MIT Press.

Norman, D.A. (1988) *The Design of Everyday Things*. New York: Doubleday/Currency.

Reisberg, B., Borenstein, J., Salob, S.P., Ferris, S.H., Franssen, E. and Georgotas, A. (1987) 'Behavioral symptoms in Alzheimer's disease: phenomenology and treatment', in *J of Clin Psychiatry*, vol. 48, S9–15.

Tyson, M.M. (1998) *The Healing Landscape: Therapeutic Outdoor Environments*. New York: McGraw Hill.

Volicer, L. and Bloom-Charette, L. (eds) (1999) *Enhancing the Quality of Life in Advanced Dementia*. New York: Taylor and Francis.

Zeisel, J. (2006) *Inquiry by Design: Environment/Behavior/Neuroscience in Architecture, Interiors, Landscape, and Planning*. New York: W.W. Norton.

Zeisel, J. and Welch, P. (1981) *Housing Designed for Families: A Summary of Research*. Cambridge, MA: Joint Center for Urban Studies for MIT and Harvard.

Zeisel, J. and Tyson, M. (1999) 'Alzheimer's treatment gardens', in Clare Cooper Marcus and Marni Barnes (eds) *Healing Gardens: Therapeutic Benefits and Design Recommendations*, New York: John Wiley & Sons.

Zeisel, J. and Raia, P. (2000) 'Non-pharmacological treatment for Alzheimer's disease: a mind-brain approach', in *American Journal of Alzheimer's Disease and Other Dementias*, vol. 15 (6), abstract published in *The Journal of Neuropsychiatry and Clinical Neurosciences*, vol. 12 (1).

Zeisel, J., Hyde, J. and Levkoff, S. (1994) 'Best practices: an environmental–behavior (E–B) model for Alzheimer special care units', in *Am J Alzheimer's Care and Related Disorders & Research*, vol. 9 (2), pp. 4–21.

Zeisel, J., Silverstein, N., Hyde, J., Levkoff, S., Lawton, M.P. and Holmes, W. (2003) 'Environmental correlates to behavioral outcomes in Alzheimer's special care units', in *The Gerontologist*, vol. 43 (5), pp. 697–711.

Part 4

Research issues: where are the research challenges and which theories and methods offer most promise?

Measuring the quality of the outdoor environment relevant to older people's lives

Takemi Sugiyama and
Catharine Ward Thompson

CHAPTER 11

Introduction

The ageing demographic in Britain and across the Western world has encouraged research into ways of improving and extending people's health into old age. This is part of a strategy not only to contain health care costs but also to improve older people's quality of life and to prolong their independence. As people age, remaining in a familiar home and neighbourhood becomes increasingly important (Laws, 1994). But, if they are to remain at home, elderly people need to be able to continue to use the wider environment, to go outdoors and, ideally, into their local neighbourhood, otherwise there is a danger that they become effectively trapped inside. Getting outside can offer benefits to people's wellbeing at a number of levels but may also present challenges and difficulties. It is thus important to investigate how and why outdoor environments affect older people's quality of life and to identify the aspects of design that help or hinder older people in using the outdoors. The research and the challenges described here are part of a larger project, entitled Inclusive Design for Getting Outdoors (I'DGO) conducted in the United Kingdom (I'DGO, 2006).

Background

In our research, quality of life is seen as a multi-layered concept which takes into account a range of factors such as health, social networks and material resources but also recognizes older people's ability to make use of the environment, adapt to it and find ways to pursue their goals. This in turn contributes to higher order needs, such as control, autonomy, self-realisation and pleasure, which are considered over-arching aspects of quality of life (Hyde *et al.*, 2003). Maintaining health into old age is an important element of quality of life and engaging in appropriate activities can help maintain physical and cognitive capabilities. There is substantial evidence that participation in physical activity has multiple benefits for older people's health (e.g., Bean *et al.*, 2004; Keysor and Jette, 2001; Mazzeo *et al.*, 1998). A physically active lifestyle can minimise the physiological changes associated with ageing and help delay or prevent the onset of common chronic diseases such as cardiovascular diseases, diabetes, arthritis and osteoporosis (Singh, 2002). Participation in regular physical activity improves older people's functional capabilities by enhancing muscle strength, aerobic capacity, balance and flexibility (Keysor and Jette, 2001). Such enhancements also help reduce the possibility of falling, which is a major cause of disability in late life (Skelton, 2001). Physical activity has also been shown to exert a positive effect on older people's cognitive functioning (e.g., Weuve et al., 2004), improve night time sleep quality (e.g., Morgan, 2003) and have a protective effect against depression (e.g., Strawbridge et al., 2002). In sum, remaining physically active contributes greatly to older people's well-being, and thus identifying and enhancing the contexts which support physically active lifestyles is important.

The outdoor environment can play a vital role here, as a context for older people to be active and maintain health. Yet, at the same time, the outdoor environment can present many challenges that make activities difficult or unpleasant for those at a stage in their lives when strength, agility and stamina are in decline. As Lawton (1986) has suggested, the combination of decreasing functional capability and barriers in the environment may act as a deterrent to outdoor activity. This trend toward inactivity is clearly shown in health statistics. In England, for instance, more than one-third of middle-aged people (aged between 35 and 54) meet the recommended level of physical activity (at least 30 minutes a day, 5 days a week), whereas only about 10 per cent of people 65 years and older meet the recommendation (Joint Health Survey Unit, 2004). Since many activities involving a moderate level of physical exercise, such as walking, bicycling and gardening, take place largely or entirely out of doors, it is important to understand what aspects of the outdoor environment encourage or inhibit such activities. Improving the quality of the outdoor environment so that getting outdoors is easy and enjoyable for older people clearly has potential to encourage active lifestyles and enhance healthy old age.

A major challenge in examining the role of the environment in relation to people's activity lies in identifying the *relevant* quality of the environment. To investigate how the environment facilitates or inhibits activity, this quality of the environment must be defined and measured. This chapter explores methods to measure this quality; we discuss basic principles in conceptualisation and propose two instruments for measurement. The chapter also reports the results from an empirical study in which the relationships between the quality of the outdoor environment, measured using these two methods, and older people's outdoor activity were examined.

Conceptualising the quality of the outdoor environment

The quality of the environment is conceptualized in this study as the extent to which it facilitates or hinders participation in activity outdoors. Thus, we call this quality the 'supportiveness' of the environment. Given that older people vary greatly in terms of lifestyle and functional capabilities, it is important to consider individual differences in the process of assessing supportiveness. The same environment may have different degrees of supportiveness for different people. In this sense, supportiveness is not a simple environmental construct, but is derived from the interaction between environmental and individual attributes. Based on this principle, we have developed two instruments to measure the supportiveness of the outdoor environment. One method takes account of individual needs and desires through a focus on the specific behaviours in which a person engages (or wishes to) in an outdoor setting, while the other looks at specific environmental attributes that have been identified empirically as relevant to people's use of the environment.

Focusing on activity

In the interaction between the environment and individual, a person may need or want to engage in certain activities, and will have a certain level of functional capability to carry them out. The environment, on the other hand, provides not only an opportunity for but also impediments to an intended activity. The choice of activity and the ease with which it can be undertaken in a setting is contingent on these personal and environmental attributes. The idea of 'personal projects analysis' was employed in the current study to capture this person–environment interaction. Personal projects refer to a set of goal-oriented, self-generated activities a person is doing or thinking of doing (Little, 1983). They range from trivial, everyday routines to ambitious, long-term endeavours. The idea of personal projects emphasizes the ecological aspects of undertaking a project by treating each project as a behaviour within a context. Each project in this concept is an interactional unit of analysis involving both personal and contextual dimensions (Little, 2000).

Kelly's (1955) personal construct theory (PCT) provides a theoretical basis for personal projects. Kelly argued that individuals have 'personal constructs', through which they construe and understand their worlds. A person interprets events and attaches meaning to them according to his or her personal constructs, which develop based on past experiences. Little's concept of personal projects shares the same constructivism with PCT. In the case of personal projects, the environment provides different people with different kinds of opportunity and demands according to their different levels of personal needs, wants and capability.

The application of personal projects in measuring the supportiveness of the environment has some advantages. First, contextual resources and constraints are likely to have a growing significance in older people's choice of activities (projects) as the general process of decline in their functional capabilities progresses. Second, the diversity of older people's lifestyles and activity patterns also makes this approach sensible because salient environmental attributes and settings may vary between them. In contrast to normative ways of measuring the quality of the environment, where criteria are fixed and are assumed to be equally salient to all the people, this idiographic method makes it possible to assess the supportiveness based on individuals' needs, wants and a relevant setting within which chosen activities are undertaken.

Focusing on environmental attributes

Some environmental attributes help people carry out activities, while others make them difficult. Research in this area has documented numerous environmental characteristics that have a bearing on people's outdoor activity patterns. The literature indicates that environmental attributes relevant to activity participation include residential density, land-use mix, street pattern (connectivity), access to shops, access to recreational facilities, quality of footpaths, aesthetics, and safety (e.g., Duncan and Mummery, 2005; Frank et al., 2005; Humpel et al.,

2004b; Li et al., 2005; Saelens et al., 2003; Sugiyama and Ward Thompson, 2005). Drawing on the findings from existing empirical research, certain instruments to assess the quality of the environment with regard to activity have been developed and reported in the literature.

There are broadly two types of instrument: one employs objective measures, while the other uses self-report or subjective measures. One example of objective measures is SPACES (Systematic Pedestrian and Cycling Environmental Scan) proposed by Pikora et al. (2002). This instrument aims to assess the suitability of neighbourhood environments for walking and cycling by combining GIS data (location of services, parks, facilities, public transport, etc), data from planning and traffic authorities (street design, traffic volume, etc.), and audit of the environment by trained observers (functionality, safety, aesthetics, etc.). A different kind of scale, NEWS (Neighborhood Environmental Walkability Scale), developed by Saelens et al. (2003), is an example of an instrument using subjective measures. This scale aims to identify the perceived adequacy of a neighbourhood area for walking as assessed by each research participant. It includes dimensions such as access to services, street pattern, availability of facilities for walking, aesthetics and safety. Humpel et al. (2004b) have reported a similar ecological scale to evaluate the quality of neighbourhood as a place to walk. Their scale consists of the following four dimensions: accessibility to facilities for walking, aesthetics, safety and weather. Both types of measures have been found to be effective in predicting people's level of activity in a neighbourhood.

An obvious advantage of objective measurement is its direct assessment of environmental attributes, less susceptible to reporting biases. An important goal in investigating environmental correlates of activity is to provide policy makers, planners and designers with helpful information to plan and implement effective environmental interventions. Findings obtained from research focused on environmental attributes can be directly translated into policy recommendations and design guidelines. In the case of subjective measurement, an advantage lies in its involvement of personal perceptions and judgement. As discussed before, for older people, the same environmental attributes may have different implications depending on a person's capabilities and other personal characteristics. Perceptual measurements such as NEWS are capable of reflecting individual differences. On this point, a recent study has shown the significance of perceived environmental factors by showing that the changes in people's perceptions of the environment account for the changes in neighbourhood walking activity (Humpel et al., 2004a). A measure proposed in this study draws on the latter principle, that is, the value in understanding individuals' perceptions of the environment, in light of the significance of individual differences among older adults.

Details of the two measures of environmental supportiveness used in this study are described below. Data were collected in the United Kingdom in 2005 to find out to what extent these measures are associated with older people's outdoor activity, and to examine the correlation between the two. It was anticipated that they would be correlated to some extent, since they aim to assess similar aspects of the outdoor environment.

Method

Data collection procedure

The methodology used in this study was a self-administered questionnaire. A hybrid sampling approach was taken to capture the diversity of the older population (people over 65 years of age) and their environmental contexts. Detailed methods of recruitment have been reported elsewhere (Sugiyama and Ward Thompson, 2007). First, 20 local authorities were selected from across Britain, based on population distribution, geographic location, residential density, type of industry, level of social deprivation (according to the national Index of Multiple Deprivation) and the balance of urban to rural areas, in an attempt to make the sample as representative as possible. The questionnaire was mailed directly to older people randomly sampled from within these local authority areas, whose names and addresses were obtained using a market research company. Thirteen organizations (housing associations and local city councils) in the same 20 local authorities also agreed to distribute the questionnaire in their sheltered housing schemes. In addition, data were collected from Chinese and Indian sub-continent ethnic groups through two translated sessions facilitated by ethnic support groups. The same questionnaire was used to collect data from all of these sample groups. The total number of responses was 335 (211 from direct mailing (response rate 10%), 102 from mailings via housing associations/city councils and 22 from translated sessions with minority ethnic groups).

Measures and Instruments

The outcome variable for this research was the quantity of outdoor activity that participants undertook. Self-reported measures of the frequency and average duration of two types of walking (walking to go to places, walking for recreation) and of other outdoor activity in a typical summer and a typical winter month were recorded. This measure primarily focuses on lifestyle activities, which may be moderately intense at times (e.g. brisk walking) but are not necessarily so, in the light of the findings indicating that this type of activity is as beneficial as structured activities (Dunn et al., 1999). From the information obtained, each participant's average time for outdoor activities (hours/week) was calculated.

As discussed earlier, one of the environmental supportiveness measures was based on personal projects (SPP). The original version of the personal projects analysis questionnaire developed by Little (1983) was simplified for this study. Participants were first asked to list outdoor activities they undertake regularly or are thinking about doing (free description). Activity was defined broadly in the questionnaire, and examples such as 'make my garden beautiful', 'walk the dog every day' and 'play bowls' were given to suggest that it can include a variety of everyday activities. Participants were then asked to evaluate each activity in terms of the extent to which the outdoor environment makes it difficult or easy to carry out, and its personal importance on a 5-point scale. The overall score (supportiveness) for each participant was calculated as a weighted mean of the difficulty/easiness ratings using the importance score as a weight (Wallenius, 1999). The instrument used in the questionnaire is shown in Table 11.1.

To measure supportiveness focusing on environmental attributes (SEA), a 26-item scale was constructed, building on published instruments which aim to measure the 'walkability' of a neighbourhood (Humpel et al., 2004a; Saelens et al., 2003). The scale incorporated the findings from eight focus group

Table 11.1 The instrument to measure environmental supportiveness based on personal projects (SPP)

Your outdoor activity (Fill in as many as you want)	How difficult does the outdoor environment make it for you to carry out the activity?					How important is this activity to you personally?				
	very difficult				very easy	not important				very important
1............	1	2	3	4	5	1	2	3	4	5
2............	1	2	3	4	5	1	2	3	4	5
3............	1	2	3	4	5	1	2	3	4	5
4............	1	2	3	4	5	1	2	3	4	5
5............	1	2	3	4	5	1	2	3	4	5
6............	1	2	3	4	5	1	2	3	4	5
7............	1	2	3	4	5	1	2	3	4	5
8............	1	2	3	4	5	1	2	3	4	5

interviews the authors conducted earlier with older people in a range of urban, suburban and rural geographical contexts, and relevant information from various design guidelines on outdoor environments (e.g., Civic Trust, 2004; Department for Transport, Local Government and the Regions, 2002). Participants were asked to rate each item on a 5-point Likert scale ranging from 'strongly disagree' to 'strongly agree'. The items in the scale are shown in Table 11.2.

Demographic variables collected included age, gender, living arrangement (own home or shelter/care home) and education (the age formal education finished – shown to be a good surrogate for socio-economic status). In addition, a measure was used to assess each participant's functional capability because it was considered likely to confound the relationship between activity and supportiveness of the environment, that is, those with better physical capabilities are likely to be more active and perceive their surroundings to be more supportive. Participants were asked to indicate the ease with which they could perform six instrumental activities of daily living (IADLs) (Jette et al., 1986). These IADLs were mostly concerned with mobility, such as walking a certain distance, climbing stairs and using public transport. For analysis purposes, the IADL scores were split into two, high and low functional capability, using the median value.

Table 11.2 Items to measure environmental supportiveness based on environmental attributes (SEA)

Item

1   There are many open spaces, parks or pedestrian-friendly routes within easy walking distance of my home.
2* There are obstacles such as busy main roads that make it difficult for me to get into the local open space.
3   The paths to get to the local open space are easy to walk on.
4   The paths to get to the local open space are enjoyable to walk through.
5   There are enough seats to rest on in the local open space.
6   There are good facilities (toilets, shelters, etc.) in the local open space.
7   It is easy to park a car next to the local open space.
8   It is easy to get to the local open space using public transport.
9   Many different activities take place in the local open space.
10 The local open space is good for children to play in.
11 The local open space is good for chatting with people.
12 The local open space is welcoming and relaxing.
13 The local open space is clean and well maintained.
14 Trees and plants in the local open space are attractive.
15 The local open space has helpful signs and information boards.
16 There is an attractive fountain or water feature in the local open space.
17* Dogs and dog fouling make the local open space unpleasant.
18* Youngsters hanging around in the local open space are a problem in my neighbourhood.
19 The local open space is mostly free from crime.
20 The local open space is safe to walk in after dark.
21 Most of the streets and paths in my neighbourhood are safe to walk after dark.
22 There are good footpaths to reach most places I need to go in my neighbourhood.
23* Steep hills and steps in my neighbourhood make it difficult to get around.
24 There are communal gardens or allotments where I could grow things near where I live.
25 There are many attractive natural features (scenery, wildlife, gardens, etc.) near where I live.
26 There is a canal, river, lake or beach that I can walk along near where I live.

* Reversely coded items

Analysis

Both supportiveness measures, SPP and SEA, were categorised into tertiles: three evenly distributed levels of supportiveness, that is, low, medium and high. Since participants' functional capability and socio-economic status might confound the relationship between environmental supportiveness and time spent in outdoor activities, the adjusted mean values of the time spent in outdoor activity were calculated, controlling for age, gender, living arrangements, education and functional capabilities. The analysis was conducted separately for SPP and SEA.

Results

The final sample size, after excluding missing and extreme cases, was 292. The characteristics of the sample are shown in Table 11.3. The table indicates that 61% of the respondents were female, and their mean age was 75. Approximately one-quarter of them continued education after 16 years old, and two-thirds of them are living in their own home. The mean amount of time spent in outdoor activity was 6.7 hours per week. This indicates that the participants in this study spend about 1 hour per day in the outdoor environment. The table shows the significant associations ($p < .001$) between functional capabilities of participants and the measures being explored, that is, supportiveness of the environment and level of outdoor activity, which suggests the importance of controlling for this variable in the analysis.

Figures 11.1 and 11.2 are scatter diagrams illustrating the relationship between each of the two measures of environmental supportiveness, as perceived by respondents, and their outdoor activity time. In both charts, it is possible to see a positive association between the supportiveness measure and activity time. Table 11.4 shows correlation coefficients between these variables and participants' functional status. The table indicates that the two environmental supportiveness measures were significantly correlated with each other, and they were also correlated with outdoor activity time and functional capability.

Figures 11.3 and 11.4 show the adjusted mean time for participants' outdoor activity by the tertile of environmental supportiveness, controlling for functional capabilities and socio-demographic variables. Both graphs show a large difference in activity time between the groups of people with low and high functional status ($p < .001$). They also show a trend in which people living in highly supportive environments tend to spend more time in outdoor activity. Although SPP showed a significant association with outdoor activity time in the previous correlation analysis, the relationship was not significant when functional capability and other variables were included in the analysis (Figure 11.3, $p = .23$). However, in the case of SEA, the association between supportiveness and outdoor activity time remained significant after adjustment (Figure 11.4, $p < .01$).

Discussion

The results of this survey show that, after controlling for respondents' functional capability and demographic variables, environmental supportiveness based on personal projects (SPP)

Table 11.3 Characteristics of the sample by functional capability category

	Functional Capability			
	Low	High	Total	p
Number	148	144	292	–
Gender (% female)	61%	61%	61%	ns
Mean age	76.4	73.7	75.0	< .01
Education (% continued after 16)	23%	28%	26%	ns
Living arrangement (% living in own home)	63%	73%	67%	ns
Mean functional capability score	2.29	4.58	3.43	< .001
Mean supportiveness score (SPP)[a]	3.34	4.41	3.90	< .001
Mean supportiveness score (SEA)[b]	2.90	3.30	3.10	< .001
Mean time for outdoor activities (hr/wk)	4.17	9.37	6.74	< .001

[a] Supportiveness based on personal projects
[b] Supportiveness based on environmental attributes

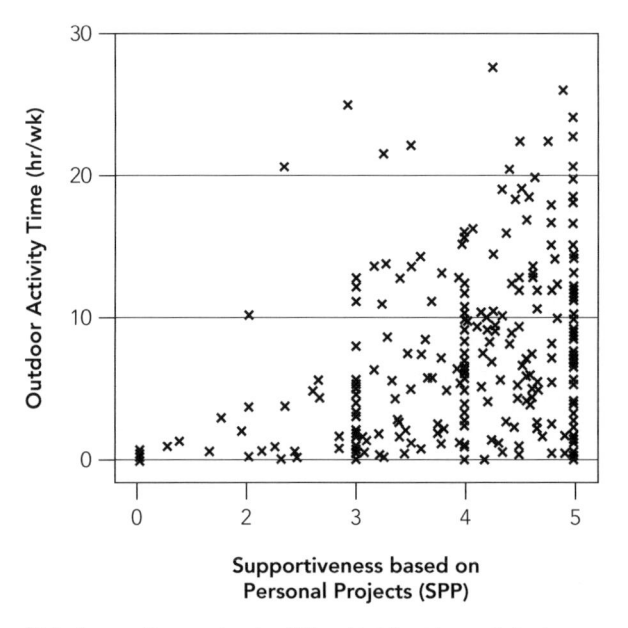

THE OUTDOOR ENVIRONMENT AND OLDER PEOPLE

11.1 Scatter diagram showing SPP and total outdoor activity time.

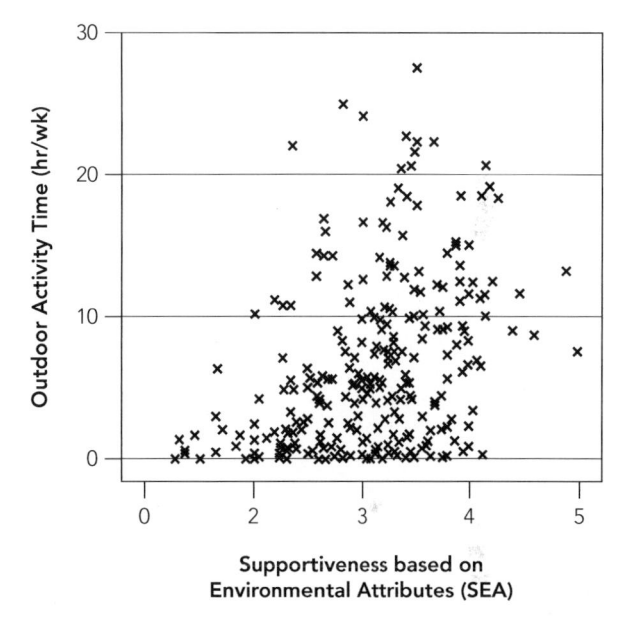

11.2 Scatter diagram showing SEA and total outdoor activity time.

Table 11.4 Correlation coefficients between SPP, SEA, outdoor activity time and functional capability

	SPP	SEA	Outdoor Activity Time
Supportiveness score (SPP)[a]	–		
Supportiveness score (SEA)[b]	0.45***	–	
Outdoor activity time (hr/wk)	0.39***	0.37***	–
Functional capability	0.65***	0.33***	0.56***

*** $p < .001$
[a] Supportiveness based on personal projects
[b] Supportiveness based on environmental attributes

did not show a significant association with outdoor activity time. One potential reason is that, in answering the personal project questions, participants may have chosen only those activities they consider readily manageable, rather than including activities that they may wish to do but have difficulty carrying out. Regardless of the condition of their outdoor environment, those who only list activities that a surrounding setting readily allows them to do will have the highest score of environmental supportiveness. Figure 11.1 clearly illustrates this trend: many participants with the highest supportiveness score (5) report little activity time. Thus, to correctly assess the quality of the environment relevant to activity using this instrument, it is important that respondents include not just activities that they currently find easy but also activities that they cannot carry out as much as they would like to, due to difficulties that may include environmental barriers. A modification of the instructions given for completing the SPP questionnaire could overcome this problem. Another reason for non-significance in the relationship between SPP and outdoor activity time may be the high dependency of this measure on participants' functional capability. As Table 11.4 shows, the SPP and functional capability scores were

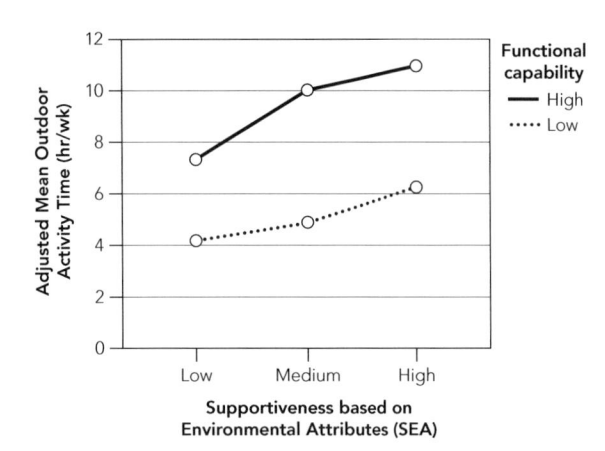

11.3 Adjusted mean outdoor activity time of the low and high functional capability groups by the level of SPP

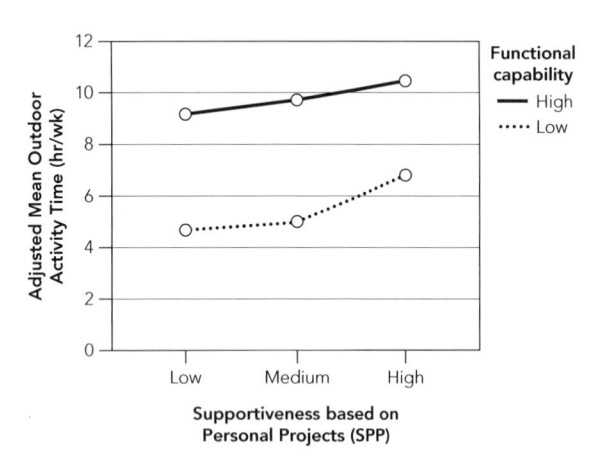

11.4 Adjusted mean outdoor activity time of the low and high functional capability groups by the level of SEA

very highly correlated ($r = 0.65$ $p< .001$). This suggests that SPP in this study measured not only the environmental quality but also individuals' functional characteristics.

Environmental supportiveness based on environmental attributes (SEA) was a significant predictor of the time spent in outdoor activity in this sample. The items employed in the scale, shown in Table 11.2, are a mixture of items related to neighbourhood environments and local open spaces in the neighbourhood. They are concerned with various aspects of these spaces, including access, comfort, pleasantness, safety and nuisance. The analysis suggests that these items are likely to be relevant to older people's participation in outdoor activity. It is notable that environmental supportiveness measured by this method was also dependent on participant's functional status to some extent. As shown in Table 11.4, the correlation between SEA and functional capability was significant ($r = 0.33$, $p< .001$). However, this is much lower than the correlation between SPP and functional capability, which means that this measure was less dependent on individuals' functional characteristics.

A significant correlation between the two supportiveness measures was observed. They were significantly correlated even when controlling for functional capability ($r = 0.32$, $p< .001$). Although the coefficient is not very high, this means that they are sharing some variance, suggesting that they are measuring similar aspects of the environment.

Conclusion

The research described here illustrates some of the challenges in finding meaningful ways to explore the relationship between the physical environment and people's levels of activity, especially in relation to the diverse characteristics of the elderly population. Yet, if there is a relationship between levels of activity and the quality of the local outdoor environment (and our findings suggest that there is), it is important for the health of the population that we understand what aspects of the outdoor environment are important, and why. We have explored the notion of 'environmental supportiveness' as a way of conceptualising the relationship between the outdoor environment and physical activity at a personal level. Drawing on theories and empirical research previously conducted, we have developed two instruments that offer possibilities for measuring the quality of the environment relevant to older people's level of activity. Both measures have demonstrated a positive association with the level of outdoor activity that older people undertake. The measure focused on environmental attributes (SEA), which assesses people's perceptions of their environment, is an effective instrument in measuring the quality of the environment relevant to older people's activity. This measure was a significant predictor of the amount of outdoor activity, regardless of people's functional capabilities, and can offer findings in a form that can be directly translated into policy recommendations and design guidelines. The measure of environmental supportiveness according to personal projects (SPP) was also examined. Although it was less robust in demonstrating this relationship distinct from people's functional capabilities, this is the first attempt to use people–environment interaction in assessing the supportiveness of the environment for physical activity. This method is potentially useful to assess the quality of the environment in terms of what it offers a particular population for whom getting outdoors poses certain kinds of challenge. It may be also employed as an exploratory tool to identify what types of behaviours people engage in, and what setting and what activities are subject to difficulties. It offers a unique way of investigating how well individuals' needs, desires and aspirations are supported by their environment and how well people cope with the environment in which they find themselves, reflecting the transactional relationship between person and environment. While methodological issues in administering the instrument need to be resolved, we suggest nonetheless that it is worthy of further exploration. It offers a person-centred approach to understanding how outdoor environments affect older people's quality of life and how aspects of that environment may help or hinder older people in using and enjoying places.

Acknowledgements

This study is part of the I'DGO (Inclusive Design for Getting Outdoors) research project, supported by the Engineering and Physical Sciences Research Council (UK). The I'DGO consortium is a partnership of Edinburgh College of Art and Heriot-Watt University, Oxford Brookes University and the University of Salford. The Sensory Trust, RICAbility, the Housing Corporation and Dementia Voice were non-academic partners in the first I'DGO project. For more information, visit www.idgo.ac.uk.

References

Bean, J.F., Vora, A. and Frontera, W.R. (2004) 'Benefits of exercise for community-dwelling older adults', in Archives of Physical Medicine & Rehabilitation, vol. 85 (7), pp. S31–S42.

Civic Trust (2004) Green flag award guidance manual. Available at www.greenflagaward.org.uk/manual (accessed July 2005).

Department for Transport, Local Government and the Regions (DTLR) (2002) Improving urban park, play areas and green spaces. Available at www.odpm.gov.uk/index.asp?id=1127724 (accessed July 2005).

Duncan, M. and Mummery, K. (2005) 'Psychosocial and environmental factors associated with physical activity among city dwellers in regional Queensland', in Preventive Medicine, vol. 40 (4), pp. 363–372.

Dunn, A.L., Marcus, B.H., Kampert, J.B., Garcia, M.E., Kohl, H.W. and Blair, N.S. (1999) 'Comparison of lifestyle and structured interventions to increase physical activity and cardiorespiratory fitness – A randomized trial', in Journal of the American Medical Association, vol. 281 (4), pp. 327–334.

Frank, L.D., Schmid, T.L., Sallis, J.F., Chapman, J. and Saelens, B.E. (2005) 'Linking objectively measured physical activity with objectively measured urban form – Findings from SMARTRAQ', in American Journal of Preventive Medicine, vol. 28 (2), pp. 117–125.

Humpel, N., Marshall, A.L., Leslie, E., Bauman, A. and Owen, N. (2004a) 'Changes in neighborhood walking are related to changes in perceptions of environmental attributes', in Annals of Behavioral Medicine, vol. 27 (1), pp. 60–67.

Humpel, N., Owen, N., Iverson, D., Leslie, E. and Bauman, A. (2004b) 'Perceived environment attributes, residential location, and walking for particular purposes', in American Journal of Preventive Medicine, vol. 26 (2), pp. 119–125.

Hyde, M., Blane, D., Higgs, P. and Wiggins, R. (2003) 'A measure of quality of life in early old age: the theory, development, and properties of a needs satisfaction model (CASP-19)', in Ageing & Mental Health, vol. 7 (3), pp. 186–194.

I'DGO (2006) About I'DGO Inclusive Design for Getting Outdoors. Available at www.idgo.ac.uk (accessed 9 October 2006).

Jette, A. M., Davies, A.R., Cleary, P.D., Calkins, D.R., Rubenstein, L.V. and Fink, A. (1986) 'The Functional Status Questionnaire: Reliability and validity when used in primary care', in Journal of General Internal Medicine, vol. 1 (3), pp. 143–149.

Joint Health Survey Unit (2004) Health Survey for England 2003. London: The Stationery Office.

Kelly, G. (1955) The Psychology of Personal Constructs. New York: Norton.

Keysor, J.J. and Jette, A.M. (2001) 'Have we oversold the benefit of late-life exercise?', in Journals of Gerontology Series A: Biological Sciences & Medical Sciences, vol. 56 (7), pp. M412–M423.

Laws, G. (1994) 'Aging, contested meanings and the built environment', in *Environment and Planning A*, vol. 26, pp. 1787–1802.

Lawton, M.P. (1986) *Environment and Aging* (2nd ed.). Albany, NY: Center for the Study of Aging.

Li, F.Z., Fisher, K.J., Brownson, R.C. and Bosworth, M. (2005) 'Multilevel modelling of built environment characteristics related to neighbourhood walking activity in older adults', in *Journal of Epidemiology & Community Health*, vol. 59 (7), pp. 558–564.

Little, B.R. (1983) 'Personal projects: A rationale and method for investigation', in *Environment & Behavior*, vol. 15 (3), pp. 273–309.

Little, B.R. (2000) 'Persons, contexts, and personal projects: Assumptive themes of a methodological transactionalism', in S. Wapner, et al. (eds) *Theoretical Perspectives in Environment–Behavior Research: Underlying assumptions, research problems, and methodologies.* New York: Plenum.

Mazzeo, R.S., Cavanagh, P., Evans, W., Fiatarone, M., Hagberg, J., McAuley, E. and Starzell, J. (1998) 'ACSM position stand: Exercise and physical activity for older adults', in *Medicine & Science in Sports & Exercise*, vol. 30 (6), pp. 992–1008.

Morgan, K. (2003) 'Daytime activity and risk factors for late-life insomnia', in *Journal of Sleep Research*, vol. 12 (3), pp. 231–238.

Pikora, T.J., Bull, F., Jamrozik, K., Knuiman, M., Giles-Corti, B. and Donovan, R. (2002) 'Developing a reliable audit instrument to measure the physical environment for physical activity', in *American Journal of Preventive Medicine*, vol. 23 (3), pp. 187–194.

Saelens, B.E., Sallis, J., Black, J. and Chen, D. (2003) 'Neighborhood-based differences in physical activity: An environment scale evaluation', in *American Journal of Public Health*, vol. 93 (9), pp. 1552–1558.

Singh, M.A.F. (2002) 'Exercise comes of age: Rationale and recommendations for a geriatric exercise prescription', in *Journals of Gerontology Series A: Biological Sciences & Medical Sciences*, vol. 57 (5), pp. M262–M282.

Skelton, D.A. (2001) 'Effects of physical activity on postural stability', in *Age & Ageing*, vol. 30 (S4), pp. 33–39.

Strawbridge, W.J., Deleger, S., Roberts, R.E. and Kaplan, G.A. (2002) 'Physical activity reduces the risk of subsequent depression for older adults', in *American Journal of Epidemiology*, vol. 156 (4), pp. 328–334.

Sugiyama, T. and Ward Thompson, C. (2005) 'Environmental support for outdoor activities and older people's Quality of Life', in *Journal of Housing for the Elderly*, vol. 19 (3/4), pp. 169–187.

Sugiyama, T. and Ward Thompson, C. (2007) 'Older people's health, outdoor activity and supportiveness of neighbourhood environments', in *Landscape and Urban Planning*, doi:10.1016/j.landurbplan.2007.04.002

Wallenius, M. (1999) 'Personal projects in everyday places: Perceived supportiveness of the environment and psychological well-being', in *Journal of Environmental Psychology*, vol. 19 (2), pp. 131–143.

Weuve, J., Kang, J.H., Manson, J.E., Breteler, M.M., Ware, J.H., Grodstein, F. (2004) 'Physical activity including walking, and cognitive function in older women', in *Journal of the American Medical Association*, vol. 292 (12), pp. 1454–1461.

Three steps to understanding restorative environments as health resources

Terry Hartig

Introduction

People commonly go to parks and other open spaces for respites from the demands of everyday life. Many societies try to ensure that their citizens have open spaces readily available for their respites, on the assumption that respites in parks will promote health. Available evidence allows confidence in that assumption, but many questions remain about just how open spaces make a difference for people's mental and physical health. Research on restorative environments provides some answers to such questions. In this chapter I discuss some of the tasks that researchers have taken on in the effort to understand restorative environments as health resources. I organize these different tasks under three steps in a sequence, proceeding from the study of discrete restorative experiences, to the study of cumulative effects of repeated restorative experiences, to the study of social ecological influences on access to and the use of places for restoration. Before going into the work at each step, however, I must set out some definitions and distinctions.

Definitions and distinctions

A respite in a park or other open space is simply time spent away from the demands of everyday life, as faced in the workplace or some other context (cf. Eden, 2001). During a respite, a person may walk, run, sit and read, watch the birds or any number of other activities. Much of the recent research on how access to open space promotes health has focused on activities, and in particular physical activities, such as walking. The basic idea behind the 'active living' research is that regular physical activity translates into physical fitness and so into reduced risk of a variety of chronic physical illnesses (e.g., Bedimo-Rung et al., 2005). Findings from this research have supported the claim that, by making open spaces readily accessible and attractive, people will more frequently engage in physical activity during their available respites (e.g., Giles-Corti et al., 2005).

As distinct from active living research, research on restorative environments focuses on particular psychological and social processes that run during a respite. Those processes do not depend on any one activity, but are instead common to a variety of activities. The processes to which I refer can be grouped under the rubric 'restoration', which I define here as the process of recovering physiological, psychological and social resources that have become diminished in efforts to meet the demands of everyday life. By resources I mean capabilities that come into play when people try to meet demands. Physiological resources include the ability to mobilize physical energy for action aimed at some demand, whether acute, as when hurrying to catch the train, or persistent, as when working hard for an extended period to meet a deadline. Psychological resources include the ability to maintain the necessary focus on some task at hand, even when noise or other distractions make it harder to concentrate. Social resources include the willingness of family, friends and co-workers to provide help of varying kinds, at home, at work and elsewhere. Because a person depletes various resources in meeting everyday demands, a potential or need for restoration arises regularly. New demands will almost certainly come along, so the person must secure adequate possibilities for restoration or risk not being able to meet those demands. Over time, a lack of adequate restoration can translate into problems with mental and physical health. This reasoning underlies studies concerned with the effects of poor sleep on health (Åkerstedt and Nilsson, 2003); restorative environments research extends it to waking activities that open onto restoration.

Restoration involves beneficial changes, but not every benefit warrants description as 'restorative'. Some benefits realised in a particular environment involve deepening or strengthening capabilities for meeting everyday demands. A person may for example become more self-reliant or self-confident, acquire new skills or gain in physical fitness. I have used the word 'instorative' to distinguish this other family of benefits from restorative benefits (Hartig et al., 1996).[1] To be sure, the possibility of relations between restorative and instorative benefits raises intriguing questions. For example, how might a restorative experience lead into an instorative experience in the same environment? Rather than pursue such questions here, I simply want to emphasise the value of the distinction as a means to discourage overly broad conceptions of restoration and its attendant benefits. Confusion about what restoration involves can hardly help efforts to understand restorative environments as health resources.

Restoration has environmental requirements; it will not occur under all circumstances. The theories that I discuss in the next section describe in some detail the environmental conditions under which restoration can proceed. Here, it suffices to say that restoration has two basic requirements. First, the environment permits restoration. While in the environment, a person can be relatively free of the demands that gave rise to the need for restoration in the first place. Second, the environment promotes restoration. Some demands are in a sense portable, and not closely tied to any one place; a person could feel troubled and ruminate over problems almost anywhere. However, some environments have features and afford activities that attract and hold attention. By drawing a person's thoughts away from demands, those features and activities can lead the person into a restorative experience and prolong it. This presence of positive features, and not only an absence of negative ones, underlies the definition of a restorative environment as an environment that promotes, not merely permits, restoration.

Earlier, I distinguished research on restorative environments from active living research, but I can also describe it as an important complement, in that it bears on some relatively undeveloped components of active living research. For example, Giles-Corti et al. (2005) found that the people they studied were more active in more attractive places. One could ask whether the characteristics of places that those people considered attractive also promoted restoration. Further, in that stress reduction and other aspects of restoration are sought-after benefits of physical activity, one could ask whether people in greater need of restoration consider physical activity in a green space more attractive than the same physical activity in a potentially less restorative environment (Staats et al., 2003; Staats and Hartig, 2004; Hartig and Staats, 2006). One could also ask whether the degree to which people realise restoration through physical activity depends on the restorative quality of the environment in which they perform it, and not solely on the activity itself (Bodin and Hartig, 2003). Some of the studies discussed in the next section offer answers in this regard.

Step 1: The study of discrete restorative experiences

Environmental psychologists and other environment-behaviour researchers have taken on several tasks in trying to explain how environments can promote restoration and, in turn, health. One of those tasks concerns discrete restorative experiences, isolated in time. The theoretical and empirical studies that I organize under this task have aimed at understanding just what happens between a person and an environment that helps restoration proceed. Because the work on restorative experiences provides points of departure for work on the tasks described in the next two sections, I treat it here as the initial step toward understanding restorative environments as health resources.

In important respects, the research on restorative environments originated in concerns for the availability of parks and open spaces. Social reformers, environmental activists and other practically minded people in nineteenth-century Europe and North America argued that access to places with natural scenery would serve the public's health, especially in the gritty urban conditions of the time. Some of those people formulated simple theories about how visits to parks would yield restorative benefits, and they used their theories in efforts to ensure that people would gain or retain opportunities to make such visits. The writings of Andrew Jackson Downing (1869), John Muir (1911) and Frederick Law Olmsted (1865/1952) exemplify the use of such theories in advocacy for public parks as means to promote the health and welfare of the public.

Although these advocates may have created and effectively applied theories about the restorative effects of natural environments, their theorizing lacked scientific grounds in the sense of well-structured, valid empirical evidence of the relations proposed. Olmsted provides a good example in this respect. Consider this quotation from his 1865 report on management of the land that eventually became known as Yosemite National Park:

> It is a scientific fact that the occasional contemplation of natural scenes of an impressive character, particularly if this contemplation occurs in connection with relief from ordinary cares, change of air and change of habits, is favorable to the health and vigor of men. . . . The want of such occasional recreation where men and women are habitually pressed by their business or household cares often results in . . . softening of the brain . . . mental and nervous excitability, moroseness, melancholy or irascibility.
>
> (1865/1952: 17)

Despite Olmsted's apparent ambition to give his statements the authority of science, I doubt that solid evidence existed to back up his assertion that lack of recreation would cause 'softening of the brain'. In broad outline his theory corresponds quite well with what today could be described as a biopsychosocial perspective on the determinants of health, but Olmsted did not venture to provide detailed descriptions of the processes through which the environment comes to affect health. Nonetheless, his writings about the restorative value of natural environments inspired the more recent theorising about restorative environments, and well-designed and carefully executed studies guided by the more recent theories have supported other of Olmsted's claims.

The authors of the two currently prominent theories about restorative environments also responded to practical concerns about access to natural environments, but they could build their arguments on bodies of empirical research that had hardly begun to develop at the time that Olmsted wrote his report. The two theories offer different perspectives on what happens during a restorative experience; they deal with different antecedents – the condition from which the person becomes restored – and they emphasise different restorative benefits.

Attention restoration theory (Kaplan and Kaplan, 1989; Kaplan, 1995) deals with the renewal of a depleted capacity for directing or focusing one's attention. According to this theory, restoration from attentional fatigue can occur when a person gains psychological distance from tasks, the pursuit of goals and the like, in which he or she routinely must direct attention (being away). Restoration is then promoted if the person can rely on effortless, interest-driven attention (fascination) in the encounter with the environment. When the person can turn loose his or her attention, so to speak, he or she can rest the cognitive mechanism that would otherwise work to inhibit attention from going to things that are more interesting than the task at hand. Further, fascination can be sustained if the person experiences the environment as coherently ordered and of substantial scope (extent). The theory also acknowledges the importance of the match between the person's inclinations at the time, the demands imposed by the environment and the environmental supports for intended activities (compatibility). A lack of compatibility can constrain being away, fascination or a sense of extent. The Kaplans argue that these four restorative factors commonly hold at high levels in natural environments, but they do not claim that only natural environments are restorative. Whether restoration takes place in a natural environment or some other environment, it becomes manifest in a renewed ability to focus and so, for example, in an improved ability to complete tasks that require concentration.

Psychoevolutionary theory (Ulrich et al., 1991; see also Ulrich, 1983) concerns stress reduction rather than attention restoration. It emphasizes the beneficial changes in physiological activity and emotions that occur as a person views a scene. For someone experiencing stress after a situation that involved challenge or threat, viewing a scene might open into restoration. This initially depends on visual characteristics of the scene that can very rapidly evoke an emotional (or, more precisely, affective) response of a general character, such as interest or fear. This response is thought to be 'hard-wired'; it does not require a conscious judgement about the scene, and indeed it can occur before a person can formulate such a judgement. The characteristics of the scene that elicit the response include gross structure, gross depth properties and some general classes of environmental content. In the case of restoration, the process would go something like this: a scene with moderate and ordered complexity, moderate depth, a focal point and natural contents such as vegetation and water rapidly evokes positive affect and holds attention, displacing or restricting negative thoughts and allowing autonomic arousal heightened by stress to sink to a more moderate level. The role of natural contents in this process has evolutionary underpinnings, according to Ulrich; humans are biologically prepared to respond rapidly and positively to environmental features that signal possibilities for survival. Restoration becomes manifest in emotions and in physiological parameters such as blood pressure, heart rate and muscle tension.

Although the two theories offer different perspectives on what happens during a discrete restorative experience, they appear to complement one another in important respects. The arousal and negative emotions characteristic of stress can occur in the absence of directed attention fatigue, while, conversely, elevated arousal and negative emotions need not always accompany attentional fatigue (Kaplan, 1995). Yet the two conditions may sometimes be related. Some researchers have discussed attentional fatigue as an after-effect of stress (Cohen, 1978), and others have treated it as a condition that makes a person more susceptible to stress (Kaplan, 1995; cf. Lepore and Evans, 1996). Thus, stress and attentional fatigue may sometimes occur alone, but in other circumstances they may have some form of reciprocal relationship or otherwise coincide. Researchers who want to study how restoration proceeds in different environments must therefore consider just what their study participants will need restoration from. That knowledge, and knowledge of the restoration process(es) thereby potentiated, guides choices regarding what to measure, when, how frequently and for how long a period.

This last point needs some elaboration, but before I provide it I should say a bit more about *where* researchers have collected measures in their efforts to understand discrete restorative experiences. The 'where' is typically not one environment, but two or more, as the approach taken has largely been experimental, comparing the effects of different environments. In their experiments, researchers have commonly referred to some basic practical concerns in deciding on environments to compare. On the one hand, they have been concerned about the circumstances of people living in cities. Most people today face their household and work demands in urban areas, where the outdoor conditions may exacerbate the need for restoration rather than promote it. On the other hand, researchers have been concerned about the psychological as well as ecological costs of the loss of natural areas to urbanization. The problem is thus twofold: nearby settings that may afford relatively good possibilities for restoration often give way to environmental conditions likely to increase restoration needs. The experimental conditions that reflect these dual concerns, then, are natural settings available to urban residents versus urban outdoor settings, such as streets and pedestrian malls, to which urban residents could otherwise resort in the absence of natural areas. Attention restoration theory and psychoevolutionary theory both predict that in such a comparison the natural environment will better promote restoration than the urban one. This prediction has been tested in a number of experiments to date, in field settings and in laboratories with video or photographic slide simulations.

The two theories make a common prediction, but it is a general one. More specific predictions differ across the theories, and those differences influence choices about the type, timing, frequency and duration of measurement. Psychoevolutionary theory supports predictions about the immediate physiological and emotional effects of viewing natural versus other kinds of environment. To capture those effects, researchers have taken multiple physiological measurements from research participants over a relatively short period of time, first before they were exposed to some kind of stressor, then while they faced the stressor (e.g., a gory film), and then immediately afterwards, as they viewed one or other of the environments being compared. Reports of emotion have usually been collected at only one or two time points, perhaps before and then after viewing the environment. Such studies have shown that looking at scenes of nature can more completely reduce blood pressure and other indicators of physiological arousal toward their pre-stressor levels within a brief span of time (4 minutes in Ulrich et al., 1991; 10 minutes in Hartig et al., 2003a). Viewing scenes of nature can also quickly evoke more positive emotions and reduce negative emotions (within 7–15 minutes, if not before; e.g. Ulrich, 1979; Hartig et al., 1999; Van den Berg et al., 2003).

For its part, attention restoration theory supports predictions about environmental effects on the performance of tasks that require directed attention. It does not, however, specify how long it should take for those effects to emerge. Researchers have used a variety of attentional measures in studies guided by this theory. Some of those measures resemble tasks that a person could do at work, such as proofreading a text for spelling and punctuation errors. Other measures have been adopted from clinical assessment procedures originally designed to detect neuropsychological problems that show up in attentional capabilities. Whatever the measure used, participants typically will not complete it until they have spent some time in one or another of the environments under study, as the act of completing the measure interrupts the restorative experience. The period in the environment typically follows a period in which the research participant does work that demands directed attention, thereby inducing some degree of attentional fatigue. After the induction of fatigue, participants have spent varying amounts of time either walking within or viewing scenes of an environment before completing the attentional measure(s). Effects on performances have not consistently emerged after 7–20 minutes (cf. Hartig et al., 1996; Laumann et al., 2003; Van den Berg et al., 2003), but they have appeared after longer periods, from 30 to 50 minutes (e.g., Hartig et al., 1991; Hartig et al., 2003a). However, the comparability of effects across different durations is hampered by the fact that studies have differed in other ways. They have used different attentional measures, different procedures for inducing attentional fatigue before the environmental 'treatment', and more. Further studies can provide a better picture of the time course of attentional restoration in different environments.

Why do researchers take interest in these details? One reason is that they may have a practical value. For example, consider the studies that show that simply looking at natural scenery for a short while can better promote restoration than looking at built features common in urban surroundings. Those findings encourage attention to a broader range of possibilities for placing natural elements in urban spaces; restoration need not only rely on large green spaces where people can spend a lot of time moving around.

Researchers have another important reason to take interest in the details of discrete restorative experiences. Knowledge of what happens, and when, during respites in different environments informs a scholarly discussion about the theories used to guide research. The theories I have discussed attribute outcomes to the operation of some mechanism within a restorative process. Yet questions remain about the correctness of those attributions. The problem is a complicated one. For example, different processes might lead to the same outcome, or the same process might produce different kinds of outcome at given points in time, or different processes may run in parallel, generating different outcomes at different times. Keeping track of the details of the various studies can help researchers understand whether the outcomes are due to the process described in theory. If a theory does not enable correct attributions of outcomes to the process in focus, then researchers must modify it or reject it.

Researchers have used one or more of three approaches to the attribution of outcomes to processes. One approach, already noted, involves comparing environments that differ in ways thought to have relevance for restoration. A second approach involves tracking the emergence of different outcomes in the environments being compared. For example, in one field experiment, my colleagues and I had our research participants complete an attentional measure before, during and after a walk in either a nature reserve or along streets in an area of medium-density urban development (Hartig et al., 2003a). By looking at change from the first to the second attentional measure, we could see that distracting aspects of the urban environment caused a distinct decline in performance, while the natural environment enabled a slight improvement (see Figure 12.1). From the walk to post-walk measure, performance improved an additional small increment in the natural environment, while it rebounded slightly, though not completely, in the urban condition. Further, by measuring blood pressure at frequent intervals in the same study, we could see

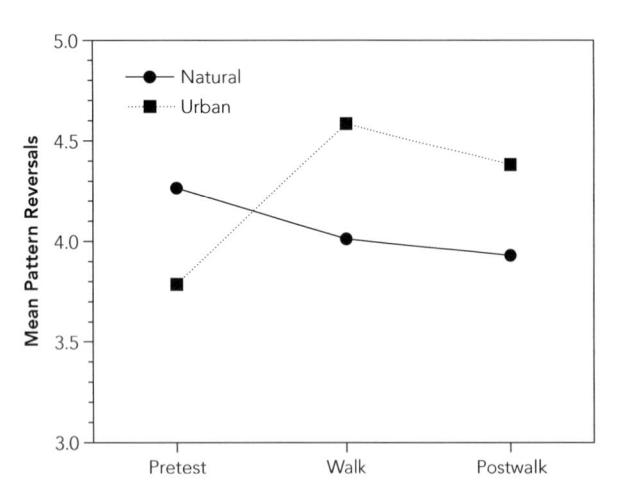

12.1 Change in attentional performance as a function of environment.

Note: The participants were to look at a 3-dimensional line drawing of a cube, called a Necker cube. A curious thing happens while one looks at a Necker cube; the side that appears to be the front suddenly becomes the back side. This is referred to as a pattern reversal. We measured the number of reversals that occurred despite the participants' efforts to concentrate on one pattern during two 30-second periods. Thus, higher values represent worse performance. Figure adapted from Hartig et al. (2003a).

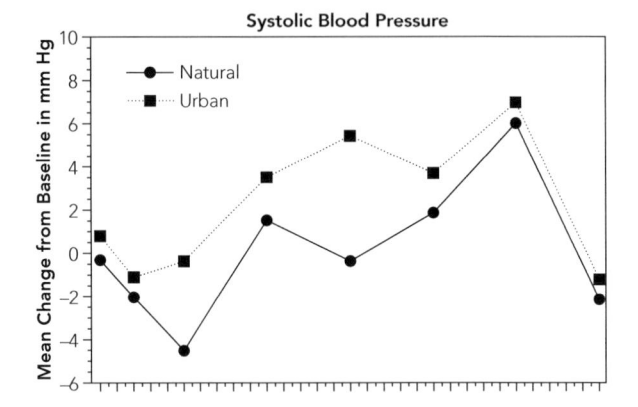

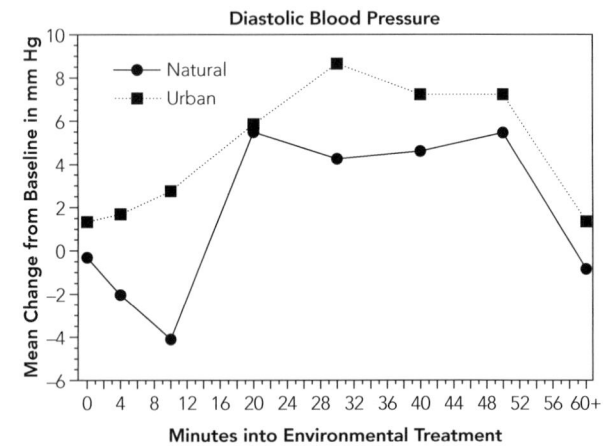

12.2 Change in systolic (top panel) and diastolic (bottom panel) blood pressure relative to baseline as a function of environment.

Note: The readings at 4 and 10 minutes occurred while the participants sat in a room with window views of trees and vegetation or in a viewless room. The readings at 20, 30, 40 and 50 minutes occurred during a walk in a nature reserve or an area of medium-density urban development. The readings at 60+ minutes occurred while the participants again sat in a room with window views of trees or in a viewless room. Figure adapted from Hartig et al. (2003a).

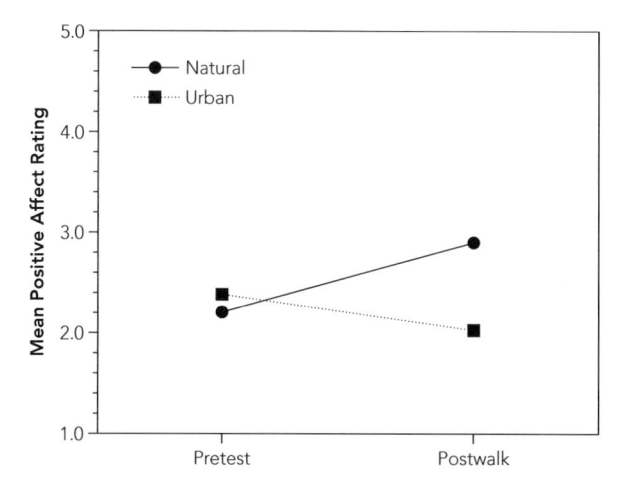

12.3 Change in self-reported positive affect as a function of environment.

Note: Scores could range from 1 to 5. Higher scores indicate greater positive affect. Figure adapted from Hartig et al. (2003a).

how physiological and attentional changes followed different paths over time in the two environments. During the walk, blood pressure initially increased in the urban condition but decreased in the nature reserve, but by the time that the participants were re-seated in the field laboratory, the difference between groups had essentially disappeared (see Figure 12.2). Yet when the participants reported on their current emotions (or affect) when back at the field laboratory after the walk, those who had walked in the nature reserve showed an increase from the pretest in positive affect, while the those who had walked in the urban surroundings showed a decline (see Figure 12.3). The picture that emerges from this and other studies is of the differential emergence of restorative benefits, some sooner and some later, and the differential dissipation of restorative benefits, some more quickly and some more slowly. The pattern of results thus speaks to the way in which different processes run over time.

A third approach to attributing outcomes to processes involves measuring people's perceptions of the environment in terms of the restorative qualities proposed in theory. Several groups of researchers have put forward measures of the restorative qualities proposed in attention restoration theory, initially for use with adults (Hartig et al., 1991; Hartig et al., 1997; Laumann et al., 2001; Herzog et al., 2003) and recently one for children (Bagot, 2004). These measures consist of a set of statements, such as 'This place is fascinating', which study participants rate on a given scale, ranging, for example, from 'not at all' to 'completely'. Valid and reliable measures of being away, fascination, extent and compatibility would help researchers address a variety of questions. For example, they could more directly assess the role of fascination in mediating an effect of environment on a capacity for directing attention. They could also assess the relative importance of the different restorative qualities in supporting restoration in given places (cf. Korpela and Hartig, 1996).

Measures of restorative qualities may eventually prove useful for practical purposes. For example, if research can establish that fascination contributes to attention restoration, then a means to measure fascination can inform the design of environments for attention restoration. One can, of course, ask whether members of the design community would find such measures useful in practice. As it stands, some have recognized the value of translating restorative environments theory into design options (Cooper Marcus and Barnes, 1999; Kaplan et al., 1998). To the extent that the value of the translation rests on the validity of the underlying theory and details of discrete restorative experiences, they are a matter of professional interest to design professionals as well as researchers.

Step 2: The study of cumulative effects of repeated restorative experiences

I have treated research on discrete restorative experiences as the initial step toward understanding restorative environments as health resources. An important next step involves studies of the cumulative effects of repeated restorative experiences. Knowing what happens in a discrete restorative experience has undeniable value, but researchers generally assume that one such experience will of itself ordinarily do little to promote lasting good health and well-being. Rather, a basic assumption underlying research on restorative environments concerns their cumulative effects: a person who accesses environments of high restorative quality during periods when restoration can occur will cumulatively realise greater restorative benefits than he or she would do by spending the time in environments of lesser restorative quality.

Three components of this assumption have important practical and methodological implications. One involves the environments to which a person has visual or physical access. The second involves those periods or respites in which restoration can occur, whether brief and in passing or of substantial duration and dedicated to the purpose of restoration. The third involves the span of time over which repeated restorative experiences can produce some cumulative effect. Taken together, these components of the 'cumulative effects assumption' have encouraged attention to people in their everyday contexts, where they would ordinarily and regularly seek out or otherwise find possibilities for restoration on a regular basis over an extended span of time.

The focus on people in their everyday contexts has meant that studies of cumulative effects have differed from studies of discrete restorative experiences in important ways. Some of those differences follow from the fact that researchers have little or no ability to exercise direct experimental control over people in their everyday contexts. For one, researchers typically cannot randomly assign people to reside, work or recreate under specific environmental circumstances for months or years at a time. Instead, they can study people who by some natural process have come to have different possibilities for accessing restorative environments in their everyday contexts. Sometimes, researchers can establish that the people under study came into different environmental conditions by an essentially random process (see Kuo and Sullivan, 2001). When researchers cannot be certain of this, however, they have to guard against the possibility that the people under study differed in relevant ways before they came into the different environments. Such 'self-selection' may pose problems even when the people initially came into the different environmental conditions by a random process, should particular kinds of people leave one or more of the environmental conditions at a disproportionate rate before or during the period covered by the study.

Similarly, in contrast to the studies of discrete restorative experiences, most studies of their cumulative effects have involved little experimenter control over key features of the environmental treatment. That is, researchers have not had control over the total amount of time their participants spent in the given environment during some period of interest; the frequency, duration and timing of their stays; what they did, with whom, when; and so on. What researchers have been able to do is to establish a time frame during which the people under study could have spent a substantial amount of time in one or another of the comparison environments. During that time, the people under study would presumably have had numerous opportunities for discrete restorative experiences of varying kinds, just as they would have frequently faced the kind of resource-depleting everyday demands that cause a need for restoration. Such an approach may make the most of the available research opportunity and can yield helpful results; however, it may not be able to address some questions, such as those concerning the frequency and duration of restorative experiences needed for cumulative benefits.

A third difference between studies of discrete restorative experiences and studies of their cumulative effects involves the outcomes of interest. The character of cumulative effects can vary widely. The basic resources restored in discrete restorative experiences come into play in most if not all aspects of daily life. Because of that, regularly having more or less effective restorative experiences can ultimately play out in many different mental, social and physical health outcomes. Those outcomes examined in studies of cumulative effects have been of substantial import, meaningful to individuals and society alike. A single restorative experience in one versus another environment could hardly be expected to show up as differences in these outcomes. For their part, studies of discrete experiences have usually measured outcomes that reflect on a hypothesised restorative process, but which in and of themselves have little lasting importance. The greater practical import of the outcomes in focus justifies the efforts required to deal with limited experimenter control and self-selection in studies of cumulative effects.

The studies of cumulative effects do share some common features with the studies of discrete restorative experiences. Notably, they have started from the same theories, attention restoration theory and psychoevolutionary theory, and they have most often operationalised the restorative quality of environments in terms of natural features. Some of them have also used measures that have been used in studies of discrete restorative experiences, to enable attribution of the outcome of ultimate concern to the restorative process thought to be at work.

People in their residential context have attracted much of the attention in studies of cumulative effects completed to date. The focus is easily understandable. Given that most people spend a large proportion of their waking as well as sleeping hours within their dwelling or in the area around it, variations in restorative quality there if anywhere could reasonably be expected to have cumulative effects. Some studies have focused on adults, notably the studies of low-income housing residents in Chicago by Kuo and Sullivan (e.g., Kuo and Sullivan, 2001; Kuo, 2001). Those residents came to occupy their housing through something like a natural lottery, with assignment to the particular dwelling units made by the public housing authority. Kuo and Sullivan grouped their study participants according to the amount of trees and other greenery around the multi-family buildings in which they lived. Thus, residents had varying opportunities for seeing trees from their windows, socialising with friends and neighbours under the trees and so on. Despite the modest 'doses' of nature involved, the residents in the greener buildings performed better on a standardised test of attention than the residents in buildings with barren surroundings. The particular test of attention they used has also been used in studies of discrete restorative experiences, but in this case it was taken as an indication of a persistent deficit in the ability to direct attention. The residents' performance on the attentional task in turn predicted their ability to manage major life issues, as well as the amount of aggressive and even violent behaviour that they had directed at members of their family. The data suggest that difficulties in managing major life issues and aggressive behaviour may stem from a weakened ability to direct attention. Kuo and Sullivan thus provided plausible evidence that repeated instances of attention restoration supported by visual and physical access to greenery can have cumulative effects on behaviours of substantial consequence to individuals, families and society.

Taylor, Kuo and Sullivan (2002) tested a similar set of hypotheses with children living in the same low-income housing area. They reported that the greener the view that 7–12-year-old girls had from their residence, the better they performed on tests of attention, and the better they could inhibit impulses and delay gratification. Again, the results suggest that repeated occasions of attention restoration can have cumulative effects on important outcomes. Other studies, using different research designs and measurement strategies, have also produced evidence of cumulative effects of restorative experiences available in the residential context, as reflected in outcomes such as residential satisfaction (Kaplan, 2001), attachment to the residence (Korpela et al., 2001), and psychological distress in children (Wells and Evans, 2003).

The workplace is another everyday context in which many people regularly and over an extended span of time find opportunities for restoration. Here, too, researchers have taken interest in possible cumulative effects of discrete restorative experiences and the influence of environmental variations on those effects. For example, Kaplan (1993) discussed the potential cumulative value of 'micro-restorative experiences' in workplaces, emphasizing that a worker might more effectively restore the cognitive resources needed for work if he or she can periodically look out of a window on to natural features such as trees and vegetation. Acknowledging the importance of taking breaks away from the desk or workstation, Kaplan nonetheless argued that brief micro-restorative experiences while at one's desk could play an important role in reducing attentional fatigue because the worker must face that immediate work setting more continuously. Results from one workplace study reported in her paper suggested that workers who had window views on to natural features had greater satisfaction with their jobs. In a second survey study, workers who had a view on to natural features gave more positive evaluations of the job and reported higher life satisfaction more generally. The report does not make clear, however, what other worker and workplace characteristics were included as statistical controls in the assessment of these relationships.

Health care settings have also drawn the attention of researchers interested in cumulative effects of restorative experiences. Hospitals, clinics and doctor's offices do not belong to the everyday life of most people, but many people can count on spending some time in such settings at some point. Ulrich's (1984) seminal study of environmental effects on recovery from surgery started from awareness of the stress and anxiety that people often face in such settings. Perhaps concerned about a threat to health, perhaps waiting for a painful procedure, a patient's unsettled state may be exacerbated by an unfamiliar environment experienced as sterile, noisy or otherwise unpleasant. Ulrich studied the records kept for patients who, after surgery, were placed in a room that had a window view of either trees or a brick wall. During the second through fifth days of their stay, those with the tree views used fewer potent pain killers than similar patients who had a view of a brick wall. Those with tree views also had shorter postoperative stays and fewer negative evaluations from nurses. This study, although modest in size, has proved influential in discussions of hospital design, perhaps because the outcomes studied are so important, to patients, staff, administrators and insurers alike.

A final example here also concerns the health care context, but the study in question was not carried out in health care settings. In the rationale for the study, Cimprich and Ronis (2003) referred to the problems that attentional fatigue can cause cancer patients, such as impairing their ability to gather information about the disease, make decisions about their treatment and follow their treatment programme. Cancer patients might cope better with their illness if they could build regular restorative experiences into their ordinary routines. To test this possibility, Cimprich and Ronis randomly assigned women with newly diagnosed breast cancer to either an intervention group or a standard care control group. The women in the intervention group committed to spending two hours a week engaged in some kind of restorative activity in a natural setting, such as gardening or visiting a scenic location. The intervention was initiated in the period between the initial diagnosis and a surgical intervention, and continued beyond the surgical intervention. Cimprich and Ronis reported on change in the capacity to direct attention from the initial measurement roughly 17 days before surgery to some 19 days after surgery. They found that the women in the natural environment intervention group performed better on a battery of attentional tests than did the women in the standard care group. This study is notable in that the participants had a fairly regular schedule of rather long, discrete restorative experiences in natural environments. As such, it differs from the other studies of cumulative effects that I have described above, for which the deliberateness, frequency and duration of such experiences was not known.

These examples suffice to give a sense of the contexts in which cumulative effects have been studied and the approaches taken to studying them. Before concluding this part of the discussion, it bears mentioning one additional respect in which studies of cumulative effects have often differed from studies of discrete restorative experiences. It involves the tradeoff made between different kinds of validity. The experiments used to study the effects of discrete restorative experiences typically enable causal claims with a high degree of internal validity; that is, they allow a good measure of confidence that it is the environmental treatment and not something else that actually caused the measured outcomes. This is owing to the characteristic features of 'true' experiments, most notably the random assignment of participants to experimental conditions, but also a relatively high degree of experimenter control over the environmental treatment and other experimental design features that can help to rule out alternative explanations for the effects found. However, one can reasonably ask how well the results of these true experiments generalise beyond the given experimental circumstances. This is an issue of external validity. Some researchers accord little generalisability to true experiments on discrete restorative experiences because of the artificiality of the experimental context and the use of university students as convenient participants. In contrast, they may especially value 'quasi-' experiments on cumulative effects of restorative experiences because they involve existing groups of 'real' people in their ordinary circumstances. These features of quasi-experiments might boost the generalizability of their findings. However, the results of quasi-experiments are in general more vulnerable to alternative explanations, or threats to internal validity, than the results of true experiments. As I have already indicated, one major threat is self-selection: people who differ in important ways, even before the environment could have affected them, either select themselves into or remain in particular comparison conditions. Surveys used to study cumulative effects may also produce highly generalizable and valuable results, but they generally do not allow causal claims because they involve measures collected at only one time point. Consistently convergent results from different kinds of experimental and non-experimental studies will, over time, allow more confidence in causal claims and the generalization of those claims, both for discrete restorative experiences and for their cumulative effects. For a discussion of validity issues and tradeoffs among different kinds of validity, see Shadish et al. (2002).

Step 3: The study of social ecological influences on restoration

From discrete restorative experiences and the cumulative effects of repeated restorative experiences, I move now to discuss social ecological influences on access to and the use of environments for restoration. In doing so, I want to illustrate how a social ecological perspective can contribute to the development of research on restorative environments. A social ecological perspective encompasses the ongoing exchange between people and environment at different levels of analysis (e.g., Catalano, 1979; Stokols, 1992). Such a perspective opens up questions about processes at work above the individual level that ultimately come to affect individuals. To see what this can mean for research, consider once again the 'cumulative effects' assumption: a person who accesses environments of high restorative quality during periods when restoration can occur will cumulatively realize greater restorative benefits than he or she would do by spending the time in environments of lesser restorative quality. Consider now some questions for social ecological research suggested by this assumption. How do processes at work above the individual level affect the average restorative quality of the environments available to the members of some populations? How do such higher-order processes affect access to environments of high restorative quality? How do they affect the periods in which restoration can occur? Thus, social ecological research in this area still concerns the cumulative effects of repeated restorative experiences, but one of its important tasks involves describing how higher-order processes affect the distribution in space and time of the environments available to people for restoration. Having described such effects, research can document their further implications for health, for populations as well as for individuals.

To my knowledge, few empirical studies of restoration and restorative environments have started from a social ecological perspective; however, the practical rationale for research on restorative environments has long had a social ecological component. In keeping with that rationale, researchers have discussed the implications of their results with regard to interventions that change the environmental conditions in which many people could spend time. Consider the following example:

> As with regular sleep, regular access to restorative environments can interrupt processes that negatively affect health and well-being in the short- and long-term. For urban populations in particular, easy pedestrian and visual access to natural settings can produce preventive benefits. Environmental strategies for health promotion that improve opportunities for restoration can offset limitations of individual-based behavioral change approaches . . ., and they complement approaches focused on preventing, eliminating, or mitigating stressor exposures.
> (Hartig et al., 2003a: 122)

Such statements follow the example set by F.L. Olmsted and his contemporaries, and they suggest how science and environmental professions can work above the individual level to improve the 'average' restorative quality of the environments that are available to members of a population, as well as their access to environments of high restorative quality.

Numerous other processes can work above the individual level to affect access to and the use of environments for restoration, for better or for worse. In the following, I will describe in some detail two studies that considered how higher-order processes might affect restoration. The two studies do not bear directly on the health resource values of open spaces, but the issues addressed do relate to the use and valuation of open spaces.

The first study focused on home-based telework as an activity that could mitigate stress from some demands, such as commuting, and yet compromise the restorative quality of the home (Hartig et al., 2007a). Two higher-order processes of interest to us in this study were the shift from an industrial to a service economy and the evolution of information and communication technologies (ICT). Together, these processes have increased possibilities for moving regular paid work from dedicated workplaces into the homes of workers. That people can work at home does not, however, mean that they will necessarily do so. My colleagues and I took the adoption of telework as a reflection on how people use their homes to cope with the demands of everyday life. In particular, we viewed ICT as a means by which people could overcome spatial and temporal constraints in their daily cycles of activity. In teleworking, they sought a better way to coordinate efforts to meet demands in different places and times, and so to mitigate sources of stress. For example, empirical research on telework and telecommuting had shown that people often telework to avoid commuting or to reconcile competing demands from childcare and paid work (e.g. Mokhtarian and Salomon, 1994; Ellison, 1999). We also knew that teleworkers sometimes come to feel isolated in their homes, tend to work longer hours and sometimes find it hard to get sufficient psychological distance from work. We went a step further by asking whether teleworkers experience less effective restoration than non-teleworkers, should the restorative quality of the home have been compromised by the entry of paid work (see also Hartig et al., 2003b).

The result of yet another higher-order process, a decision by the Swedish government to decentralize one of its agencies, created an opportunity to address this question. When the Swedish National Energy Administration (*Statens Energimyndighet*; STEM) relocated from Stockholm to another city about 100 km (62 miles) away, management offered staff the option to do part of their work as telework. This reflected awareness that employees living in Stockholm would not willingly move their residences. Management drew up an agreement that allowed employees to spend up to three workdays per week working at home. Prior to the move, STEM employees did not have the option to work at home. Thus, the relocation of the STEM offices involved both a major intervention in geographical fundamentals of the employees' cycles of activity and an explicit allowance for telework as a means to cope with concomitant changes.

From responses to a survey conducted about one year after STEM began its move, we found that employees who worked at home one or more days each week frequently cited a long commute and parental responsibilities as reasons for teleworking. They also reported considerable temporal, mental and spatial overlap of work and non-work life, with only spatial overlap reduced by using a separate room for their work at home. In general, the higher the perceived overlap, the more negatively it was evaluated. Teleworking appeared to be associated with restoration, conditional on gender; of those who teleworked, women reported less, and men more, effective restoration than their counterparts among non-teleworkers. The results from the STEM study thus suggest that teleworking as a strategy for coping does not serve all equally well, and that some workers may suffer a constraint of restoration with the entry of paid work into the home (see also Lundberg and Lindfors, 2002). Just who benefits and who loses may itself reflect on higher-order processes that influence the assignment of responsibility for domestic work to women versus men (Michelson, 2000).

How then might the experience of teleworkers relate to the use and valuation of open spaces? For one, the study results suggest that a reduction in the restorative quality of an environment, especially one relied on for restoration, can prove harmful in the long run. This could hold for green spaces as well as for the home. For another, teleworkers may use greenspaces to compensate for a reduction in the restorative quality of the home. Ahrentzen (1989, 1990) reported results of interviews with homeworkers that suggest that neighbourhood green spaces become increasingly important for people who start to work at home. She cited one respondent who said that during breaks from work it was important to have a quiet walk outside the home-cum-workplace. About 20% of her sample said that having quiet walking conditions nearby had become more important to them since they started to work at home. Given this, one might ask about the number of people who work at home and the associated demand for nearby green spaces for walking during working hours. Early projections of widespread teleworking, especially as a substitute for commuting, have largely gone unfulfilled (Salomon, 1998). At the end of the 1990s, only about 8% of workers in Sweden regularly worked at a location other than the ordinary workplace during their regularly scheduled working hours (Vilhelmson and Thulin, 2001). Such a figure is not negligible, however, and higher-order processes could work to increase it, as through increasing gasoline prices, for example.

The second study that I will describe here bears on each of the effects of higher-order processes that I referred to earlier: change in the average restorative quality of available environments, change in access to environments of high restorative quality and change in the periods in which restoration can occur. The study focused on unseasonably cool weather as a likely constraint on restorative outdoor activities during the summer months in Sweden (Hartig et al., 2007b). We hypothesised that, during relatively cool summer months, reduced participation in outdoor activities would be reflected in a higher incidence of health problems in which stress has an etiological role. To test our hypothesis, we used official time-series data for the dispensation of selective serotonin reuptake inhibitors (SSRIs), a category of drugs that includes the most widely prescribed antidepressants, such as Prozac. The data came from the pharmacy system coupled to the Swedish national health care system, and they represent the number of doses dispensed to the Swedish population as a whole on a monthly basis, broken down for women and men separately. We could thus look at how the total dispensation of SSRIs varied with monthly mean temperature over a period of several years.

In addition to weather, a higher-order process of interest to us in this study was the development of legislation. Swedes have long imparted special significance to the summer as a season for leisure and recreational activities outdoors, perhaps because they live in a land with long, dark winters. From the 1930s into the late 1970s, this positive regard for the summer became manifest in vacation legislation. In an opinion from a legislative committee, issued in connection with early debate over a legally mandated summer vacation period, the authors wrote, 'It is obviously a strong desire that vacation will take place at such a time of year that it provides the greatest restoration. For the greatest number the summer appears to be the most appropriate time' (Andra Lagutskottet, 1953: 11). However, this was a controversial issue; many actors in Swedish society participated in the debate, and they voiced starkly different opinions about how much vacation workers should get and how it should be disposed over the year. In the end, the advocates for a long summer vacation won. The vacation law now requires that employers allow their employees four consecutive weeks of paid vacation during the period 1 June–31 August (Semesterlagen 1977: 480; see Ericson and Gustafsson, 1977). Surveys indicate that Swedes consistently exercise their right, and that they take a large part of their vacation during July, the month with the highest mean temperature (Statens Offentliga Utredningar, 1988).

By specifying the period during which workers could leave their workplaces for a long restorative respite, the vacation law did more than ensure that they would have enough time for restoration; it also aimed to improve their access to environments of relatively high restorative quality. However, no law can ensure that any given four-week summer period will be warm enough for people to enjoy spending time sunbathing, swimming in the sea, walking in the mountains and so on. We reasoned that unseasonably cool summer temperatures would deter the choice of activities that would otherwise aid restoration from stress induced by chronic role strains and other demands. This could work during the vacation period as well as during the remainder of the summer, when Swedes also value spending time outdoors in warm weather.

We had several reasons to expect that the constraint of restoration by cool summer weather would become manifest in the use of antidepressants. Those reasons include the documented link between stress and depression (Hammen, 2005), and the fact that outdoor activities such as walking can mitigate depression (Craft and Perna, 2004). We also knew that one strategy for treating depression seeks to engage a psychological process much like that described in restorative environments theory. Depressed people are encouraged to engage in pleasant activities such as taking a walk and looking at beautiful scenery, which can distract them from self-contemplation or rumination until their mood has lifted, opening the way for efforts to resolve the reasons for depressed mood (Lewinsohn and Libet, 1972; Nolen-Hoeksema et al., 1993).

Rumination may prolong depressive episodes to a greater degree among women than among men, which could help to account for the greater prevalence of depression among women (Nolen-Hoeksema, 1987). The tendency toward rumination may be compounded by daily demands, particularly among the parents of young children. Childcare appears to impose a relatively greater amount of demands on mothers (Lundberg, 1996), and so a relatively greater reduction in access to distracting activities. We thus hypothesised that the association between summer temperatures and the dispensation of SSRIs would be stronger among women; in relatively cool weather they would have fewer options for distraction from reasons for depression, in particular those embedded in the settings from which they would otherwise escape to the outdoors, such as role demands in the home.

In our statistical tests, we looked at each of the summer months separately, to see if the dispensation of SSRIs co-varied with average temperature over the period 1991–98. For example, the average monthly temperature for July during this period varied between roughly 15° C (in 1993) and 20° C (in 1994). The time-series analyses controlled for a number of possible influences on dispensation other than temperature for the given month. One of these was the increase in acceptance of SSRIs as the drug-of-choice for depression, as reflected in a strong upward trend. Another was the fact of broader seasonal influences on the incidence of depression. Given our hypothesis of gender differences in the 'dose response', we did separate analyses for women and men.

After taking out variation in SSRI dispensation due to trends, change in temperature over the course of the whole year and our other control variables, we found that the results provided some support for our hypotheses. When the June of a given year in the period had a relatively cool temperature, the dispensation in SSRIs was slightly higher than one would expect, but not reliably higher in a statistical sense. Similarly, we did not find a statistically significant association between mean monthly temperature and dispensation of SSRIs for the month of August. However, for the month of July, when most Swedes take a substantial part of their summer vacation, the association was statistically significant. To take July of 1998 as an example, by considering the number of women living in Sweden during that month and the number of them apparently taking SSRIs, we could use the results of the time-series analysis to estimate that some 761 fewer women would have used SSRIs for each one-degree increase in temperature for that month. The corresponding number for men would have been 251. Furthermore, the number as a percentage of the total number using the drugs would have been greater for women than for men. Thus, in line with our expectations, the 'dose response' to temperature appeared to be greater in women.

And what does this study have to say about the use and valuation of open spaces? It provides another perspective on their health resource value for populations by pointing out a possible consequence of restrictions on their use. Many people work to ensure or improve access to open spaces by arguing that they promote health. In their work, they often refer to scientific literature on the benefits of access. They might also point to the harm that can be done when access is lost or the quality of the experience is diminished; however, without empirical examples to cite, the argument is circumstantial, based on inferences from knowledge of the benefits of access. This study provides an empirical example that speaks more directly to the ways in which constrained access to outdoor activities can negatively affect mental health in a population. The activities in question include many of those in which people participate in their local parks and other open spaces, weather permitting.

The two studies I have just described illustrate how diverse higher-order processes can affect opportunities to access and use different environments for restoration. They also show how quite different methods can be used to address questions about the influence of higher-order processes on restoration and, in turn, health. And, of course, they have their limitations, much like the studies on discrete restorative experiences and their cumulative effects. In the study of teleworkers, we collected data at only one time point, so we cannot say to what degree the teleworkers and non-teleworkers showed more or less effective restoration after versus before they started teleworking. We could only use control variables in our statistical analyses to try and rule out differences that might have existed between the groups before STEM moved its main office. In the study of depression and unseasonable summer weather, my colleagues and I built on psychological theory in articulating the research question, but we worked with aggregate data for the entire Swedish population. On the basis of those data we cannot say, for example, just how much particular outdoor activities of particular individuals varied across summer months with high and low temperatures. We could only assume, on the basis of other research (e.g., Statistiska centralbyrån, 2004), that Swedes did indeed spend less time engaged in recreational activities outdoors during relatively cool summer months. It bears repeating here: consistently convergent results from different kinds of study will over time allow more confidence in causal claims and generalization of those claims.

Concluding comments

In this chapter I have referred to research on restorative environments as one source of answers to questions about how parks and open spaces can promote mental and physical health. Studies of restorative benefits realised in discrete experiences; studies of the cumulative effects of repeated restorative experiences; studies concerned with social ecological influences on access to and the use of places for restoration – findings at these three steps can serve individual and public health when translated into practical measures. Although I have focused here on research tasks undertaken at each of these steps, I do not mean to imply that they are the only tasks in need of attention. Nor do I mean to imply that there is some essential chronological order in which the different tasks must be taken on. The kind of knowledge generated has been applied on a more intuitive basis for many decades, but the research area itself is a relatively young one. At each of the steps, many questions remain to be investigated, and work on them can proceed in tandem with work on the tasks at other steps. How can we distinguish restorative benefits from instorative benefits (cf. Herzog et al., 1997)? What built settings in urban surroundings might afford discrete restorative experiences as beneficial as those that people find in natural surroundings? For what kinds of people do particular kinds of park and open space best promote restoration over the long run (cf. Van den Berg and Ter Heijne, 2005)? How does the distribution of restorative quality in everyday environments vary as a function of socioeconomic status, and how does that distribution change as socioeconomic inequalities widen? As researchers address new questions and continue to work with the tasks described here, one has good reason to hope for better arguments regarding the health benefits of open spaces and the value of making them broadly accessible.

Note

1 Although of recent origin, the word 'instorative' does have etymological grounds. Just as 'restorative' means 'of or relating to restoration' when appended to 'benefit', the word 'instorative' means 'of or relating to instoration'. For its part, 'instoration' is a simple transformation of the word 'instauration', an obsolete meaning of which is 'an act of instituting or establishing something'. At the risk of confusing matters, I should mention that 'instauration' also has the meaning 'restoration'. The word 'restoration' itself comes from the Latin 'restauration'. The word 'instore' exists in English, but it is also described as obsolete. It means 'furnish' or 'provide' (compare with 'store'). In any case, all of these words have a common Latin root (for details, see Websters Third New International Dictionary of the English Language, Unabridged, 1981).

References

Ahrentzen, S. (1989) 'A place of peace, prospect, and . . . a P.C.: The home as office', in *Journal of Architectural and Planning Research*, vol. 6, pp. 271–288.

Ahrentzen, S. (1990) 'Managing conflict by managing boundaries: How professional homeworkers cope with multiple roles at home', in *Environment and Behavior*, vol. 22, pp. 723–752.

Åkerstedt, T. and Nilsson, P.M. (2003) 'Sleep as restitution: An introduction', in *Journal of Internal Medicine*, vol. 254, pp. 6–12.

Andra Lagutskottet (1953). *Utlåtande i anledning av väckta motioner om viss ändring i 12 § lagen om semester* [Report in connection with motions raised concerning change in §12 of the vacation law] (Utlåtande Nr. 20). Stockholm: Sveriges Riksdag.

Bagot, K.L. (2004) 'Perceived Restorative Components: A scale for children', in *Children, Youth and Environments*, vol. 14 (1), pp. 120–140.

Bedimo-Rung, A.L., Mowen, A.J. and Cohen, D.A. (2005) 'The significance of parks to physical activity and public health: A conceptual model', in *American Journal of Preventive Medicine*, vol. 28 (Supplement 2), pp. 159–168.

Bodin, M. and Hartig, T. (2003) 'Does the outdoor environment matter for psychological restoration gained through running?', in *Psychology of Sport and Exercise*, vol. 4, pp. 141–153.

Catalano, R. (1979) *Health, Behavior and the Community: An ecological perspective*. New York: Pergamon.

Cimprich, B. (1993) 'Development of an intervention to restore attention in cancer patients', in *Cancer Nursing*, vol. 16, pp. 83–92.

Cimprich, B. and Ronis, D.L. (2003) 'An environmental intervention to restore attention in women with newly diagnosed breast cancer', in *Cancer Nursing*, vol. 26, pp. 284–292.

Cohen, S. (1978) 'Environmental load and the allocation of attention', in A. Baum, J.E. Singer and S. Valins (eds) *Advances in Environmental Psychology: Vol. 1*. Hillsdale, NJ: Lawrence Erlbaum, pp. 1–29.

Cooper Marcus, C. and Barnes, M. (1999) *Healing Gardens: Therapeutic Benefits and Design Recommendations*. New York: Wiley.

Craft, L.L. and Perna, F.M. (2004) 'The benefits of exercise for the clinically depressed', in *Primary Care Companion Journal of Clinical Psychiatry*, vol. 6, pp. 104–111.

Downing, A.J. (1869) 'A talk about public parks and gardens', in G.W. Curtis (ed.) *Rural Essays by A. J. Downing*. New York: George A. Leavitt, pp. 138–146.

Eden, D. (2001) 'Vacations and other respites: Studying stress on and off the job', in *International Review of Industrial and Organizational Psychology*, vol. 16, pp. 121–146.

Ellison, N.B. (1999) 'Social impacts: New perspectives on telework', in *Social Science Computer Review*, vol. 17, pp. 338–356.

Ericson, B. and Gustafsson, S. (1977) *Nya semesterlagen* [The new vacation law]. Stockholm: Tidens förlag.

Giles-Corti, B., Broomhall, M.H., Knuiman, M., Collins, C., Douglas, K., Ng, K., Lange, A. and Donovan, R.J. (2005) 'Increasing walking: How important is distance to, attractiveness, and size of public open space?', in *American Journal of Preventive Medicine*, vol. 28 (Supplement 2), pp. 159–168.

Hammen, C. (2005) 'Stress and depression', in *Annual Review of Clinical Psychology*, vol. 1, pp. 293–319.

Hartig, T. (2004) 'Restorative environments', in C. Spielberger (ed.) *Encyclopedia of Applied Psychology*, vol. 3, pp. 273–279. San Diego, CA: Academic Press.

Hartig, T. and Staats, H. (2006) 'The need for psychological restoration as a determinant of environmental preferences', in *Journal of Environmental Psychology*, vol. 26, pp. 215–226.

Hartig, T., Mang, M. and Evans, G.W. (1991) 'Restorative effects of natural environment experience', in *Environment and Behavior*, vol. 23, pp. 3–26.

Hartig, T., Johansson, G. and Kylin, C. (2003b) 'Residence in the social ecology of stress and restoration', in *Journal of Social Issues*, vol. 59, pp. 611–636.

Hartig, T., Kylin, C. and Johansson, G. (2007a) 'The telework tradeoff: Stress mitigation vs. constrained restoration', in *Applied Psychology: An International Review*, vol. 56, pp. 231–253.

Hartig, T., Catalano, R. and Ong, M. (2007b) 'Cold summer weather, constrained restoration, and increased use of anti-depressants in Sweden', in *Journal of Environmental Psychology*, vol. 27, pp. 107–116.

Hartig, T., Korpela, K., Evans, G.W. and Gärling, T. (1997) 'A measure of restorative quality in environments', in *Scandinavian Housing & Planning Research*, vol. 14, pp. 175–194.

Hartig, T., Nyberg, L., Nisson, L-G. and Gärling, T. (1999) 'Testing for mood congruent recall with environmentally induced mood', in *Journal of Environmental Psychology*, vol. 19, pp. 353–367.

Hartig, T., Book, A., Garvill, J., Olsson, T. and Gärling, T. (1996) 'Environmental influences on psychological restoration', in *Scandinavian Journal of Psychology*, vol. 37, pp. 378–393.

Hartig, T., Evans, G.W., Jammer, L.D., Davis, D.S. and Gärling, T. (2003a) 'Tracking restoration in natural and urban field settings', in *Journal of Environmental Psychology*, vol. 23, pp. 109–123.

Herzog, T.R., Maguire, C.P. and Nebel, M.B. (2003) 'Assessing the restorative components of environments', in *Journal of Environmental Psychology*, vol. 23, pp. 159–170.

Herzog, T.R., Black, A.M., Fountaine, K.A. and Knotts, D.J. (1997) 'Reflection and attentional recovery as distinctive benefits of restorative environments', in *Journal of Environmental Psychology*, vol. 17, pp. 165–170.

Kaplan, S. (1995) 'The restorative benefits of nature: Toward an integrative framework', in *Journal of Environmental Psychology*, vol. 15, pp. 169–182.

Kaplan, R. (1993) 'The role of nature in the context of the workplace', in *Landscape and Urban Planning*, vol. 26, pp. 193–201.

Kaplan, R. (2001) 'The nature of the view from home: Psychological benefits', in *Environment and Behavior*, vol. 33, pp. 507–542.

Kaplan, R. and Kaplan, S. (1989) *The experience of nature: A psychological perspective*. New York: Cambridge University Press.

Kaplan, R., Kaplan, S. and Ryan, R.L. (1998) *With People in Mind: Design and Management of Everyday Nature*. Washington DC: Island Press.

Korpela, K. and Hartig, T. (1996) 'Restorative qualities of favorite places', in *Journal of Environmental Psychology*, vol. 16, pp. 221–233.

Korpela, K. M., Hartig, T., Kaiser, F.G. and Fuhrer, U. (2001) 'Restorative experience and self-regulation in favorite places', in *Environment and Behavior*, vol. 33, pp. 572–589.

Kuo, F.E. (2001) 'Coping with poverty: Impacts of environment and attention in the inner city', in *Environment and Behavior*, vol. 33, pp. 5–34.

Kuo, F.E. and Sullivan, W.C. (2001) 'Aggression and violence in the inner city: Effects of environment via mental fatigue', in *Environment and Behavior*, vol. 33, pp. 543–571.

Laumann, K., Gärling, T. and Stormark, K.M. (2001) 'Rating scale measures of restorative components of environments', in *Journal of Environmental Psychology*, vol. 21, pp. 31–44.

Laumann, K. et al. (2003) 'Selective attention and heart rate responses to natural and urban environments', in *Journal of Environmental Psychology*, vol. 23 (2), pp. 125–134.

Lepore, S.J. and Evans, G.W. (1996) 'Coping with multiple stressors in the environment', in M. Zeidner and N.S. Endler (eds) *Handbook of Coping: Theory, Research, Applications*. New York: Wiley, pp. 350–377.

Lewinsohn, P.M. and Libet, J. (1972) 'Pleasant events, activity schedules, and depressions', in *Journal of Abnormal Psychology*, vol. 79, pp. 291–295.

Lundberg, U. (1996) 'Influence of paid and unpaid work on psychophysiological stress responses of men and women', in *Journal of Occupational Health Psychology*, vol. 1, pp. 117–130.

Lundberg, U. and Lindfors, P. (2002) 'Psychophysiological reactions to telework in female and male white-collar workers', in *Journal of Occupational Health Psychology*, vol. 7, pp. 354–364.

Michelson, H., Bolund, C., Nilsson, B. and Brandberg, Y. (2000) 'Health-related quality of life measured by the EORTC QLQ-C30: Reference values from a large sample of the Swedish population', in *Acta Oncologica*, vol. 39 (4), pp. 477–484.

Mokhtarian, P.L. and Salomon, I. (1994) 'Modeling the choice of telecommuting: Setting the context', in *Environment and Planning A*, vol. 26, pp. 749–766.

Muir, J. (1911) *Our National Parks*. Boston, MA: Houghton-Mifflin.

Nolen-Hoeksema, S. (1987) 'Sex differences in unipolar depression: Evidence and theory', in *Psychological Bulletin*, vol. 101, pp. 259–282.

Nolen-Hoeksema, S., Morrow, J. and Fredrickson, B.L. (1993) 'Response styles and the duration of episodes of depressed mood', in *Journal of Abnormal Psychology*, vol. 102, pp. 20–28.

Olmsted, F.L. (1865/1952) 'The Yosemite Valley and the Mariposa Big Trees: A preliminary report: With an introductory note by Laura Wood Roper', in *Landscape Architecture*, vol. 43, pp. 12–25.

Salomon, I. (1998) 'Technological change and social forecasting: The case of telecommuting as a travel substitute', in *Transportation Research C*, vol. 6, pp. 17–45.

Shadish, W.R., Cook, T.D. and Campbell, D.T. (2002) *Experimental and Quasi-experimental Designs for Generalised Causal Inference*. Boston, MA: Houghton Mifflin.

Staats, H. and Hartig, T. (2004) 'Alone or with a friend: A social context for psychological restoration and environmental preferences', in *Journal of Environmental Psychology*, vol. 24, pp. 199–211.

Staats, H., Kieviet, A. and Hartig, T. (2003) 'Where to recover from attentional fatigue: An expectancy-value analysis of environmental preference', in *Journal of Environmental Psychology*, vol. 23, pp. 147–157.

Statens Offentliga Utredningar (1988) *Om semester* [About vacation] (SOU 1988: 54). Stockholm: Arbetsmarknadsdepartementet.

Statistiska centralbyrån (2004) *Fritid 1976–2002* [Leisure activities 1976–2002] (Levnadsförhållanden Rapport nr. 103). Stockholm: Author.

Stokols, D. (1992) 'Establishing and maintaining healthy environments: Toward a social ecology of health promotion', in *American Psychologist*, vol. 47, pp. 6–22.

Taylor, A.F., Kuo, F.E. and Sullivan W.C. (2002) 'Views of nature and self-discipline: Evidence from inner city children', in *Journal of Environmental Psychology*, vol. 22, pp. 49–63.

Ulrich, R.S. (1979). 'Visual landscapes and psychological well-being', in *Landscape Research*, vol. 4, pp. 17–23.

Ulrich, R.S. (1983) 'Aesthetic and affective response to natural environment', in I. Altman and J.F. Wohlwill (eds) *Behavior and the Natural Environment*. New York: Plenum, pp. 85–125.

Ulrich, R.S. (1984) 'View through a window may influence recovery from surgery', in *Science*, vol. 224, pp. 420–421.

Ulrich, R.S., Simons, R.F., Losito, B.D., Fiorito, E., Miles, M.A. and Zelson, M. (1991) 'Stress recovery during exposure to natural and urban environments', in *Journal of Environmental Psychology*, vol. 11, pp. 201–230.

Van den Berg, A.E. and Ter Heijne, M. (2005) 'Fear versus fascination: An exploration of emotional responses to natural threats', in *Journal of Environmental Psychology*, vol. 25, pp. 261–272.

Van den Berg, A.E., Koole, S.L. and Van der Wulp, N.Y. (2003) 'Environmental preference and restoration: (How) are they related?', in *Journal of Environmental Psychology*, vol. 23, pp. 135–146.

Vilhelmson, B. and Thulin, E. (2001) 'Is regular work at fixed places fading away? The development of ICT-based and travel-based modes of work in Sweden', in *Environment and Planning A*, vol. 33, pp. 1015–1029.

Wells, N.M. and Evans, G.W. (2003) 'Nearby nature: A buffer of life stress among rural children', in *Environment and Behavior*, vol. 35, pp. 311–330.

On quality of life, analysis and evidence-based belief

Peter A. Aspinall

Preamble

This chapter relates to two different aspects of research in OPENspace, one of which is specific and the other more general. The specific topic is quality of life, which is increasingly seen as a key criterion or outcome measure of much social research in this area. Indeed, in the health, environment and inclusive design/planning policy and practice context, it is often the principal driver underpinning research initiatives. But how might quality of life be assessed? The first part of the chapter illustrates how one methodology (conjoint analysis) can be used to derive the preferences or importances people attach to different situations, hence giving a link to their quality of life.

The second section of the chapter centres around more general research issues which have arisen from conversations and observations at research conferences in relation to landscape architecture and environmental design. These suggest:

a) There appears to be a consensus that the research we carry out should be evidence based.
b) At the same time there is little use of what is nowadays a large range of both qualitative and quantitative research methodologies.
c) There are wide-ranging views and not a little confusion as to what counts as research. This is evident from the range of papers presented at conferences.
d) There is some unease expressed about the perceived need to 'prove things', which I understand to mean the need to apply more scientific criteria or methods to beliefs which are assumed to be already shared by the research community.

This second general section is in two parts. The first part begins with point (b) and demonstrates how two recent and flexible quantitative methods, which appear to be underused by researchers, can illuminate the type of data we collect. The second part brings points (a), (c) and (d) together in an unconventional Bayesian approach to illuminate, particularly in controversial circumstances, what is believable from typical research findings.

PETER A. ASPINALL

Assessing quality of life for visually impaired people (conjoint analysis)

In seeking a theoretical framework for quality of life, the OPENspace research centre has focused on a social rather than a medical model. While the latter is associated with the absence of illness, the social model embraces ideas such as self-efficacy, perceived control, autonomy and independence, and stress, in addition to more predictable dimensions of pleasure and satisfaction, and seems more appropriate for our research agenda (Hyde et al., 2003; Little, 1998). However, in research focused on studying the quality of life of visually impaired people, there has been to date a simpler approach (Mangione et al., 1992). The definition of quality of life (or more specifically 'vision related quality of life') is the degree to which a person's vision prevents them carrying out a range of daily tasks (e.g. reading, shopping, recognizing faces). While conventional, vision-related quality of life questionnaires explore this (Mangione et al., 2001; Massof, 1995), they do not pursue the relative priority or utility of these tasks; yet the practical consequences of vision loss to a person's quality of life are influenced by the priority given to different tasks which the person finds difficult to carry out. This is a key issue across many health-related fields, both in the allocation of limited resources and also in developing recommendations for rehabilitation or coping strategies for people (NICE, 2004). The UK National Institute for Clinical Excellence (NICE) has recommended that studies about the value or utility of health care interventions should not use rating scales but instead methods such as discrete choice or conjoint analysis. All the methods recommended by NICE are characterised by a person making a *relative choice* between alternative situations from which importance or value can be derived.

The study described here involved the use of Choice Based Conjoint with Hierarchical Bayesian analysis for individual importances (or utilities), to understand the values people attach to different degrees of task difficulty (Johnson, 2001).

Study Group

Two groups of visually impaired people attending the Princess Alexandra Eye Pavilion in Edinburgh were participants in the study. The inclusion criteria for the study were:

a) People (N=108) with glaucoma (i.e. differing degrees of visual field loss) but with relatively good central vision.
b) People (N=122) with macular degeneration (i.e. with differing degrees of central visual loss).

All participants were given a range of clinical, visual and quality of life assessments. However, as the focus of this chapter is on methods, only brief mention will be made of the broader context of the study.

Conjoint analysis is useful in preference studies because it enables us to make a choice between alternatives when several items of information are presented together. This has three advantages. First, the presentation is natural and closer to real-world choices we face when we select, decide, evaluate or buy something. Second, it removes the problem inherent in many research studies or decisional programmes, whereby overall choice is determined by the results of a combination of choices across its separate elements. Conjoint analysis is more likely to reveal realistic and integrated choices than such additive models, that is, the whole is more than the sum of its parts (Tversky and Kahneman, 1982). Third, there is considerable evidence from psychological studies that people are better at making relative choices than they are at making absolute judgements, and that relative choice judgements have greater reliability and validity for preference studies (Tversky and Kahneman, 1982).

While the origins of conjoint analysis are in academic psychology, it is market science which is one of its most prolific users (Johnson, 2001). The researcher is asked to identify a set of key attributes, for example cost, which can be present at different levels (e.g. different prices). Combinations of these attributes are presented in a paired (or multiple) comparison task to a respondent. Conjoint analysis has been used to predict preference (i.e. utilities or part 'worths') for perfumes, golf balls, health treatments, transport and so on. The technique also contains within it a validity check, that is, the model predicts user response for a particular choice task which can be compared with the actual choices people make.

Consider a typical conjoint task related to the purchase of golf balls. Suppose the attributes for choice are *price* (which has three values or levels), the *distance* the ball flies (which has three distances), and *brand* (two levels – a well-known versus

QUALITY OF LIFE AND EVIDENCE-BASED BELIEF

unknown brand name). From a relatively small number of paired comparison questions, involving different levels across all the attributes, it is possible to assign utilities to all attributes and levels and through the market simulator (which is part of the software program), to see the trade-off between attributes. For example, what extra price would people pay for a ball which flies farther, or for a known over an unknown brand?

In the study of people with visual impairment, patients received a 64-item questionnaire based on focus groups with glaucoma patients, and, using factor analysis, five principal attributes were identified which were the basis of the conjoint analysis study. These were:

- reading,
- getting about outside,
- darkness or glare,
- household chores,
- bumping into things from the side.

All these situations had presented potential problems for people with glaucoma. Each attribute was described in terms of three levels of difficulty they might present, that is, 'no difficulty', 'a few problems' or 'a lot of problems'. Because the potential permutations of possible questions is large (3^5), the conjoint software selects a smaller set of orthogonal combinations for presentation. In this case, 15 paired comparison judgements (see Figure 13.1) were all that was needed per person to generate the utilities for all combinations of attribute and level. (In practice there were four questionnaires, each with a different set of 15 orthogonal questions produced by the Sawtooth program (Sawtooth Software Inc, www.sawtoothsoftware.com). The order of presentation of each questionnaire was rotated as each person was tested.) In addition to the 15 paired comparisons, which were presented to the participants, a further two 'hold out' cards were used for the validity check – this is where the model predicts a response not used in its formation, which can then be compared with the actual responses made. In the present case, a sample size of 50 participants was calculated as sufficient (Orme, 1998).

Each person in the study was presented with 15 paired comparisons of the type shown in Figure 13.1 and asked the following question. 'If you were one of these two people shown, which of the two would you think was in the worse state of health?' People rated the task as understandable and do-able. (For people in the macular degeneration group with very poor vision, we had a reduced version of the questionnaire. Interestingly, the rank order of importance for the reduced version was the same as that in the longer form.)

Figure 13.2 shows the utilities for all attribute/level combinations in decreasing order of importance, from left to right. The range of the histogram for an attribute indicates its importance. Therefore 'central vision' (reading) followed by 'outdoor mobility' emerge as the two most important attributes. Note that ranges can be compared across attributes so, for example, a difference between no problems and a few problems with reading is more important than all the difficulties associated with darkness and glare.

Key findings

The key findings can be briefly summarised as follows:

a) *For people with glaucoma*

  i) There was no evidence of subgroups in the sample with differing priorities (as defined by demographic, clinical or visual variables).
  ii) While glaucoma is associated with a peripheral visual field loss, it was concerns over central vision (reading) which emerged from analysis as most important for people, with getting about outside as next in importance (see Figure 13.2).

Choice-based Conjoint	
PERSON 1	PERSON 2
A FEW problems with reading	NO problems with reading
A LOT of problems getting outside	A FEW problems getting outside
NO problems with darkness or glare	A FEW problems with darkness and glare

13.1 The quality of life equivalent for visually impaired people, showing three of the five attributes. Which of the two people is in the worse state of health?

Choice-based conjoint analysis: Utility values from logit analysis

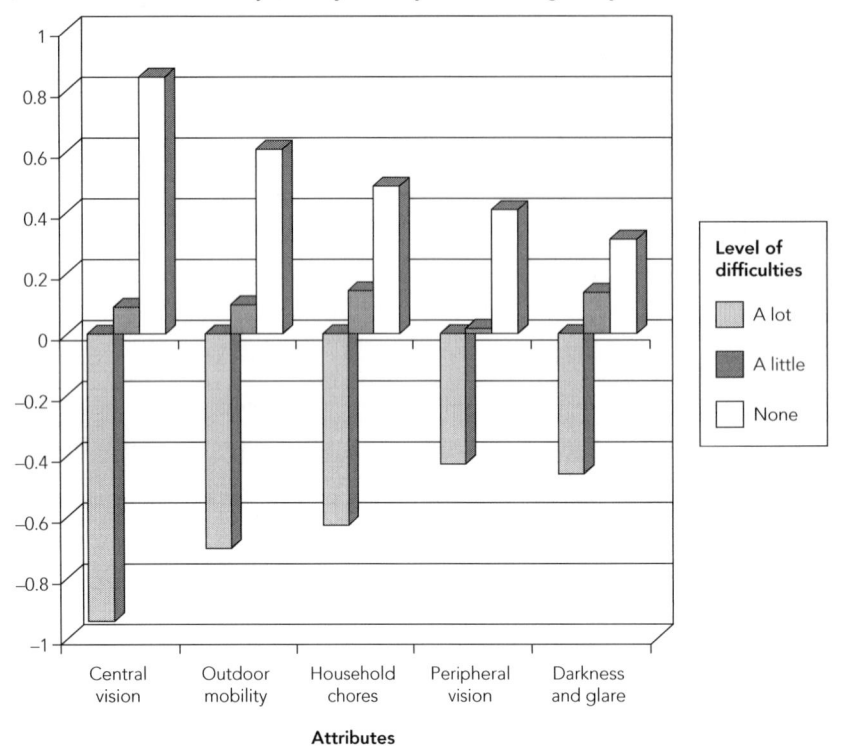

13.2 The logit analysis with importance (range) decreasing from left to right.

iii) The problem most frequently reported by patients from focus groups and questionnaires (i.e. glare) was NOT the most important under conjoint analysis.
iv) With increase in visual field loss, the relative importance of central vision increases.
v) With increase in visual field loss, the relative importance of getting about outside decreases (see Figure 13.3).
vi) Contrast sensitivity is the only measure linked to the frequently reported problems with lighting and glare.
viii) The conjoint analysis did not correlate with other questionnaire data or with the personal state of the individual.

b) *For people with age-related macular degeneration*

i) There were two subgroups of people in the sample – those most concerned with reading and central vision and those most concerned with getting about outside (see Figure 13.4).

These maps were produced from individual utilities generated by the Hierarchical Bayes program, which is part of the Sawtooth software. In other words, it is possible to give utility values for every individual in the sample as well as the average values as plotted in Figure 13.2.

ii) For most tasks and attitudes assessed, the loss in quality of life was most significant between moderate and severe forms of the disease. This underlines the value of intervention to prevent people reaching the moderate state of the condition (see Figure 13.3).
iii) Conjoint analysis findings did not relate to those from focus groups, or conventional questionnaires.

iv) Validity checks on the conjoint analysis showed the model predicting choice (on combinations of attributes the model had not previously seen) at 80% and 92% for the two hold-out cards.

The value of this technique in the context of quality of life for people with visual impairment is that it facilitates better decision-making about investment in rehabilitation and assistive tools for such people compared with conventional methods. Choice-based conjoint analysis has thus provided new insights into people's assessments of vision-related quality of life. The conjoint validity checks are high, indicating the power of the model to predict evaluations of task ability not present in the model. We are left with the dilemma as to why the different measures of quality of life are yielding differing results. As the methodology is novel in the field, we have no current answer to this. Three studies are in print – the two studies reported here (Aspinall et al., 2005; Aspinall et al., 2006) and also an application of the method to predict younger people's housing preferences (Leishman et al., 2004). OPENspace researchers are currently using this technique to explore older people's preferences in relation to accessible outdoor environments.

Useful and flexible quantitative methods

Illustrative data from studies on older people's travel diaries, and on people's use of woodland visits, is used here to demonstrate the value of some under-used quantitative methods. The two analytical techniques selected are non-parametric forms of

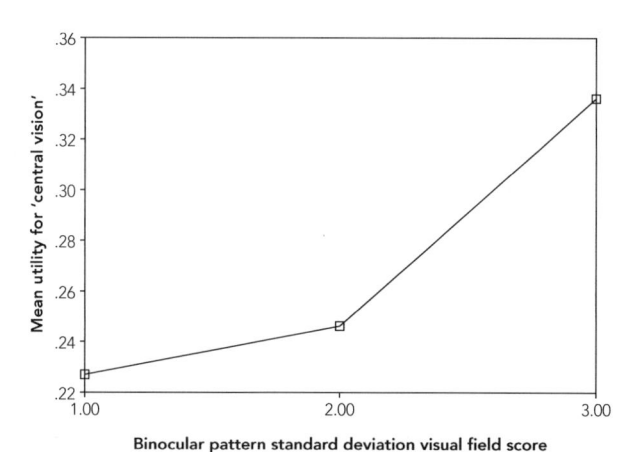

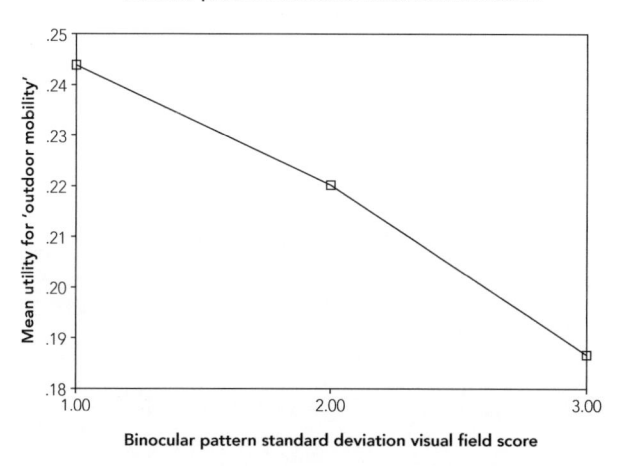

13.3 Importance (utility) of reading (central vision) (top graph) and outdoor mobility (bottom graph) changing in different ways with decreasing visual field (horizontal axis).

analysis, that is, useable on data which don't correspond to a normal distribution, and therefore require fewer constraints on the form of data collected. It is important to appreciate that, in addition to conventional forms of data gathering and measurement, there are much looser forms that can be equally valuable. If, for example, you think it reasonable in a given context for a person to be asked whether A is more or less preferred, important, valued or likely to occur than B, then there are many sophisticated analytical tools available. On the other hand, if you don't find these questions reasonable then there is a genuine problem over assessment or quantification! Within OPENspace, we have found analysis by Conjoint Analysis, AnswerTree (SPSS Inc, www.spss.com) and Latent Class (Statistical Innovations Inc, www.statisticalinnovations.com) all relevant and useful and in addition are examining Rasch analysis (Rummlab Pty Inc, see www.rummlab.com.au), which has commonalities with the group.

a) AnswerTree or regression by pictures – an example from travel diaries with visually impaired people

The SPSS module 'AnswerTree' is an analytic technique that presents regressional analysis (i.e. a way of predicting one variable from several others) in a highly visual and flexible way. It has several advantages. First, it is an exploratory decisional analysis which gives flexibility to the user to follow different strategies. AnswerTree produces a hierarchical tree structure with the target variable (i.e. the one we are interested in predicting) at the top of the 'tree'. Independent variables (i.e. the ones we wish to use to predict the target) are arranged at a series of levels within the tree. Those at the top of the tree, near the target variable, are the most significant discriminators of the target variable, while those lower down the tree are less important. The analysis therefore provides the significant discriminators, as in conventional regression, but also provides the optimal sequence in which they can be used in any decisional problem. Users can manipulate the model to provide alternative criteria at different 'nodes' of the tree and receive feedback on the impact of this change to the tree as a whole. This feature of local control should appeal to researchers and designers with special concerns or needs. Because the analysis is non-parametric it is able to cope with a variety of non-normal distributions and nominal data. In addition to the optimal basis for a sequence of decisional nodes, it identifies the most efficient criterion point on any variable for critical paths at any node, as illustrated in the example below.

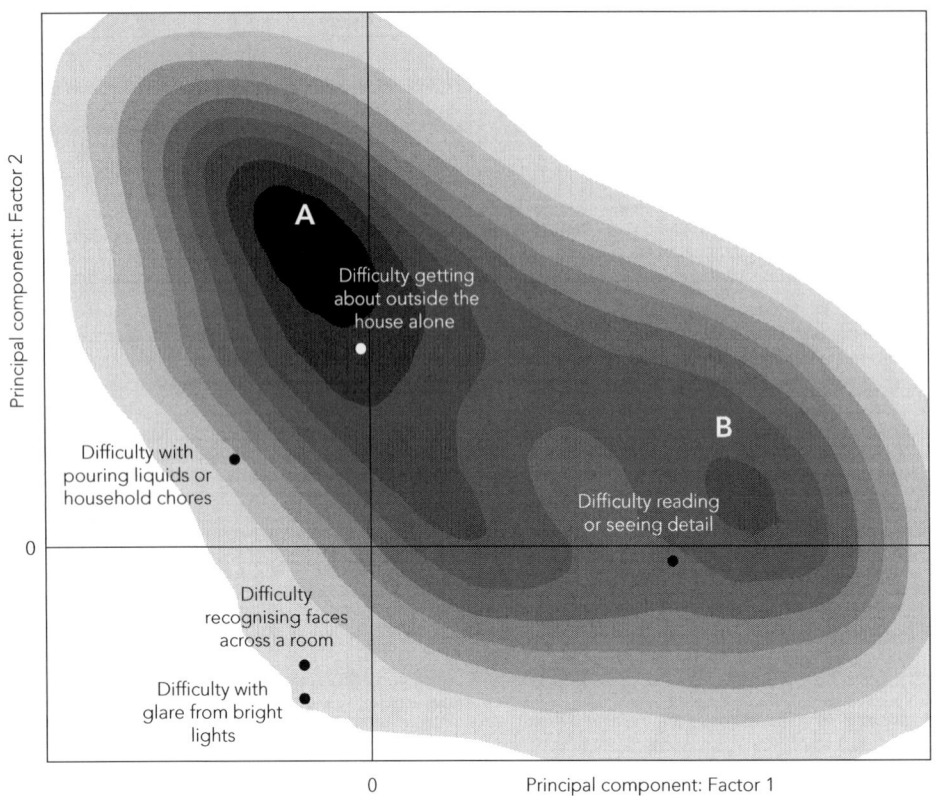

13.4 Individual utilities mapped on two main attributes (denser areas represent a greater number of people's preference vectors).

In the example which follows, a decisional tree is presented using a binary data split at each junction of the tree (the Classification and Regression Tree, CART, software option). The evaluation of each tree is given by the model accuracy in per cent or the misclassification index which follows it. The OPENspace group has included AnswerTrees in two publications – one showing that childhood experience predicted adult frequency of visits to woodlands (Ward Thompson et al., 2004); and a second showing that emotional problems were central to older people's dissatisfaction with vision services (Hill and Aspinall, 2004).

In order to clarify the logic behind the 'trees' and the decision strategy which follows it, a simple version of a tree is 'grown' here, step by step.

Step 0

In the example chosen, 69 people attending the low vision clinic at the Princess Alexandra Eye Pavilion in Edinburgh were asked to keep travel diaries over a two-week period. The group had varying degrees of vision loss and were mainly elderly. One of the target research questions was to look at the average distance of a walking trip. In the top box of Figure 13.5, the numbers of people walking less than 0.5 mile, 0.5–1.0 mile and more than 1.0 mile is shown. This is the baseline data. If we had to anticipate the walking distance of anyone in the sample with no further evidence we would say <0.5 miles because that is the most frequently occurring value. However, we would be wrong for (27+13) or 40 people. So 40/69 or 58% becomes our initial error rate.

Step 1

Let us grow the tree one level.

Figure 13.5 shows that the best overall predictor of walking distance is the age of the person and the optimal discrimination point on the age scale is 77 years. Notice how the numbers in the branches change. Of the 29 people in the total sample who are walking <0.5 miles, we now have 6 who are less than 77 years while the other 23 who are over 77 years are in the right-hand branch of the tree. The risk estimate (or error rate) given in the program output is now 44% instead of the initial 58%. So one branch of the tree has reduced the error by 14%. It is already interesting to see that age, rather than vision in the better eye, is the best predictor of mobility in this group. Conventionally, vision in the better eye has been usually regarded as most closely linked to quality of life (Rubin et al., 2000), of which mobility is one facet.

Step 2

If the tree is grown by two further levels we get the structure in Figure 13.6.

The two main predictors at the first level are now 'walking in the town centre', which was a response to the question 'How

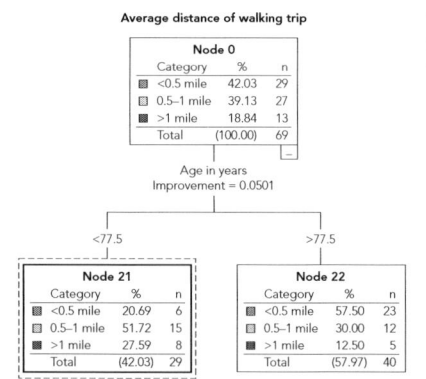

13.5 First level of AnswerTree regression, showing age as key variable.

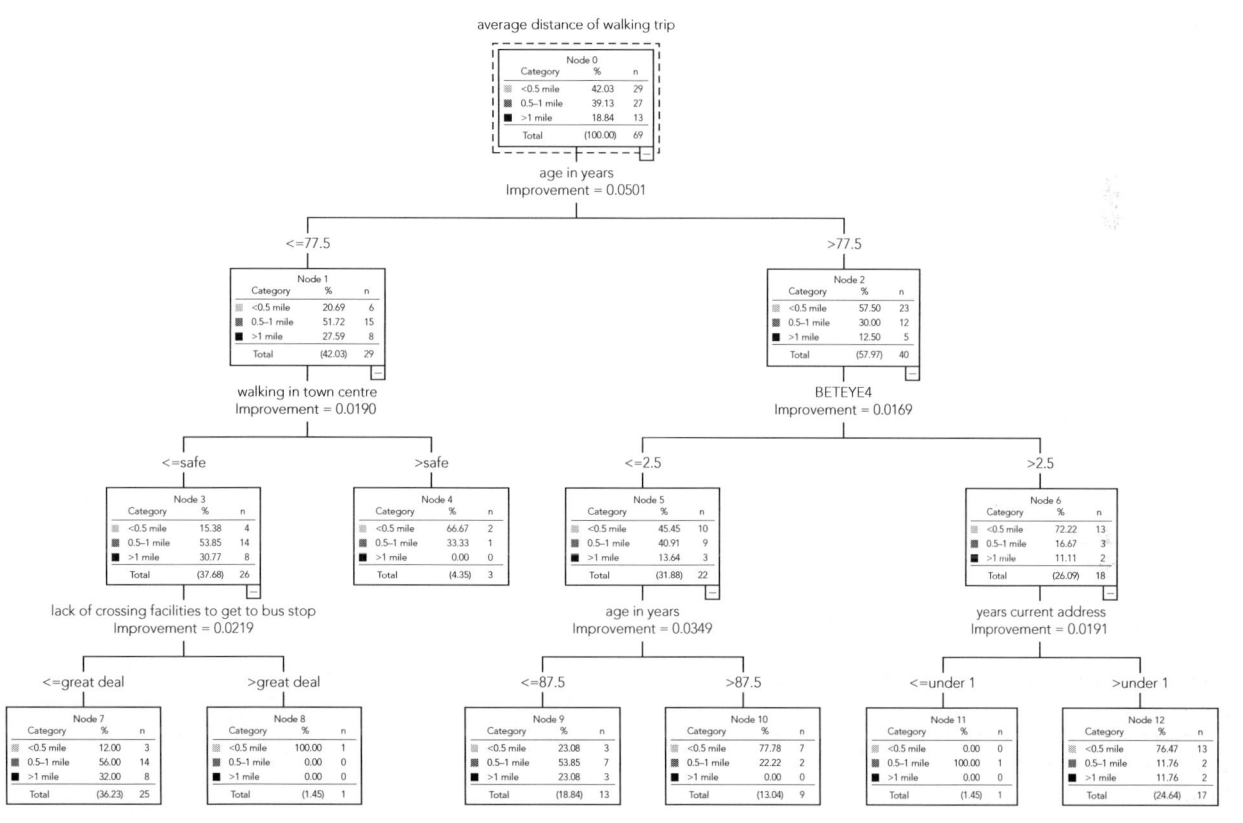

13.6 The full regression tree to 4 levels.

safe do you feel walking in the town centre?' and vision in the better eye. It is therefore an aspect of vision and feelings of safety which next predict walking distance. If we then move down to the final level, there are three predictors. One is the availability of crossing facilities to get to the bus stop, a second is years a person has lived at their current address, and the third is age. Note that there is no logical contradiction in a variable such as age appearing at more than one point in the tree. It happens to be the best discriminator at this point, but notice that the age criterion value is now 87 years. The risk assessment or error rate for this final tree is now 34%.

The regression analysis has identified the best predictors of walking distance and prioritised their influence in a way that makes it easier to identify interventional strategies.

b) Latent class analysis (or finding hidden subgroups) – an example from a study of people's use of local woodlands

In the previous study, the purpose of analysis was to find the variables which were predicting the target variable. However, in this section, the purpose is to discover whether there are subgroups of people in the study who are similar enough to be members of one subgroup (or cluster) and different enough from those in another cluster. Each cluster has therefore a particular pattern or profile of response across the variables of a study. As the subheading suggests, these profiles of clusters are hidden and can only be revealed through analysis. The analysis package we have used to achieve this is Latent Class (Statistical Innovations Inc, see www.statisticalinnovations.com). Identifying subgroups of people who are characterised in different ways or who behave in different ways, or whose preferences are different is important in user studies of environmental design and in rehabilitation studies. Researchers need to know whether they are dealing with one homogeneous group or several different ones who may have different characteristics, needs or behaviours.

The example illustrated here is taken from OPENspace data on the frequency of visits to woodlands in the central lowlands of Scotland. Other analysis of this data has been published elsewhere (Ward Thompson et al., 2004; Ward Thompson et al., 2005) and latent class is used here only for illustration.

For purpose of clarity, four variables only were placed in the cluster analysis. These were frequency of study participants' visits to woodlands, their childhood frequency of woodland visits, age and gender. The research question was aimed at discovering whether there was one homogeneous group of people in the sample or more. With the exception of gender, all the variables were split into three categories.

In the first part of the output from analysis, the software indicates how many clusters it has found from the minimum value of a particular statistic. In this case it identified two clusters. Table 13.1 shows significance or 'p' values for the variables across the two clusters. In this case, the clusters were different on all the four variables with the strongest differences, as shown by the minimum p values, being in frequency of visit and in childhood experience.

The next output, in Table 13.2, shows in the first row that 69% of the total group of people are in cluster 1, with 30% in cluster 2.

The values under the cluster columns are probabilities of being in any age, gender and so on category, given a person is in cluster 1 or 2. So, for instance, given a person is in cluster 1, the probability of being in the Age <34 group is 0.2024 or 20.24%.

Table 13.1 Measures of significance of difference between clusters for each indicator variable

Models for Indicators	Cluster 1	Cluster 2	Wald	p-value	R^2
Age	0.1509	−0.1509	4.8560	0.028	0.0508
Gender	−0.8971	0.8971	6.8561	0.0089	0.1335
Frequency of woodland visits	−0.3910	0.3910	11.7239	0.00062	0.1570
Frequency of childhood woodland visits	−0.4048	0.4048	10.1996	0.0014	0.1894

Table 13.2 Probability of indicator variable given cluster membership

		Cluster 1	Cluster 2
Cluster Size		0.6959	0.3041
Indicators			
Age	< 34 years	0.2024	0.3921
	35–54 years	0.3968	0.3951
	> 55 years	0.4008	0.2128
	Mean	4.8812	4.0706
Gender	Male	0.5880	0.1918
	Female	0.4120	0.8082
	Mean	1.4120	1.8082
Frequency of woodland visits	Daily or weekly	0.4848	0.1428
	Monthly	0.3000	0.2603
	Yearly or none	0.2152	0.5969
	Mean	2.6497	3.6896
Frequency of childhood woodland visits	Daily	0.4128	0.1141
	Weekly	0.3779	0.2347
	Monthly or less	0.2092	0.6513
	Mean	2.0037	3.1826

QUALITY OF LIFE AND EVIDENCE-BASED BELIEF

Table 13.3 Probability of cluster membership given indicator variable

		Cluster 1	Cluster 2
Overall Probability		0.6959	0.3041
Indicators			
Age	<34 years	0.5406	0.4594
	35–54 years	0.6972	0.3028
	>55 years	0.8120	0.1880
Gender	Male	0.8761	0.1239
	Female	0.5378	0.4622
Frequency of woodland visits	Daily or weekly	0.8972	0.1028
	Monthly	0.6988	0.3012
	Yearly or none	0.4622	0.5378
Frequency of childhood woodland visits	Daily	0.9002	0.0998
	Weekly	0.7769	0.2231
	Monthly or less	0.4260	0.5740

We can see from the table that:

a) cluster 1 is associated with more frequent visits to woodlands and that cluster 2 is associated with less frequent visits;
b) cluster 1 has more older people, while cluster 2 has more younger people;
c) cluster 2 has many more females than males, whereas the numbers are more even in cluster 1;
d) childhood experience of woodlands is more likely in cluster 1 and much less likely in cluster 2.

Table 13.3 presents similar data linking indicators to clusters, but now the probabilities are inverted. Here we have the probabilities of being in a cluster given a particular indicator category. So, for example, given a person is in the Age <34 group, the probability of being in cluster 1 is 0.5406 or 54%.

The table shows:

a) if you are in the older age group >55, the probability you are in cluster 1 is 0.812 whereas it is only 0.1880 that you are in cluster 2;
b) if you visit woodlands frequently you are much more likely to be in cluster 1 than cluster 2;
c) if you are male you are much more likely to be in cluster 1;
d) if you visited woodlands frequently as a child you are much more likely to be in cluster 1.

Finally, in Table 13.4, a profile is given of the probability of cluster membership given any combination of indicator variables. (Note only part of the full table is illustrated.)

It is possible, therefore, to get a probability of membership of either of these clusters, representing two distinguishable groups of people, for any person with any profile across these variables.

Research: what is it and are its findings believable? An introduction to Bayesian inference

Conference discussions suggest that the criteria or boundaries for what is or is not research (while clear in some disciplines) are not at all clear in landscape architecture, urban design and environmental design-related fields. There appear to be two sources of confusion. The first is over the question 'when is designing research and when is it not research but, rather, applied good practice?' Clarification of this distinction would be especially helpful to the inclusive design field, however this is something designers need to address. The second source of confusion is over the credibility of research findings in situations where there is a discrepancy, or relevance gap, between experiential professional opinion and scientific evidence. It is this second source of confusion, concerning the credibility of research findings, and their usefulness in relation to evidence-based belief, on which I would like to comment.

In well established scientific disciplines, the corpus of what is considered to be known is more secure than in emerging disciplines. Put crudely, more people believe a larger set of common propositions about their subject with higher degrees of certainty, and this common consensus of belief operates as a reference level for comparison with what is likely to be new or challenging (i.e. research). However, in practice-based fields such as ours, where there is an understandable mix of personal

Table 13.4 Probability of cluster membership for any combination of indicator variables

AGE	GENDER	FREQ	CHILDEXP	Modal	Cluster 1	Cluster 2
<34	Male	daily/weekly	daily	1	0.9780	0.0220
<34	Male	daily/weekly	weekly	1	0.9519	0.0481
<34	Male	daily/weekly	monthly/less	1	0.7980	0.2020
<34	Male	monthly	daily	1	0.9379	0.0621
<34	Male	monthly	weekly	1	0.8705	0.1295
<34	Male	monthly	monthly/less	1	0.5728	0.4272
<34	Male	yearly/none	daily	1	0.8253	0.1747
<34	Male	yearly/none	weekly	1	0.6777	0.3223
<34	Male	yearly/none	monthly/less	2	0.2955	0.7045
<34	Female	daily/weekly	daily	1	0.8809	0.1191
<34	Female	daily/weekly	weekly	1	0.7670	0.2330
<34	Female	daily/weekly	monthly/less	2	0.3964	0.6036
<34	Female	monthly	daily	1	0.7152	0.2848
<34	Female	monthly	weekly	1	0.5277	0.4723
<34	Female	monthly	monthly/less	2	0.1823	0.8177
<34	Female	yearly/none	daily	2	0.4399	0.5601
<34	Female	yearly/none	weekly	2	0.2590	0.7410
<34	Female	yearly/none	monthly/less	2	0.0652	0.9348
35–54	Male	daily/weekly	daily	1	0.9886	0.0114
35–54	Male	daily/weekly	weekly	1	0.9747	0.0253
35–54	Male	daily/weekly	monthly/less	1	0.8849	0.1151
35–54	Male	monthly	daily	1	0.9671	0.0329
35–54	Male	monthly	weekly	1	0.9290	0.0710
35–54	Male	monthly	monthly/less	1	0.7229	0.2771
35–54	Male	yearly/none	daily	1	0.9019	0.0981
35–54	Male	yearly/none	weekly	1	0.8036	0.1964
35–54	Male	yearly/none	monthly/less	2	0.4494	0.5506

knowledge from experience and scientific findings, this reference base receives less consent and seems more complex. There is, as a consequence, a wide variation in the degree to which many academics and professionals hold differing degrees of belief about many aspects of their subject. This is the context within which many professionals working in the field of designing, planning and managing the outdoor environment openly acknowledge the need for a greater emphasis on making their discipline evidence based. However, there is a problem. Scientific ways of underpinning belief by evidence (e.g. conventional hypothesis testing) sometimes seem of little use, or even alien to professional ways of knowing, practising or guiding design.

This is where Bayesian inference is relevant as a more radical philosophy offering a different rationale for evidence-based belief. Furthermore, for some disciplines such as medicine, where doctors have to make decisions about individual patients, there have been lead editorials in the *British Medical Journal* recommending that Bayesian ideas provide helpful diagnostic insights and may be used to guide judgement (Freedman, 1996). As landscape architects and other designers are also faced with decisions about individual environmental situations for which the evidence might be problematic, there are clear parallels. What is presented here is called naïve Bayes, and that is appropriate because it is the underlying ideas that are important rather than any detail of sophisticated computation on how they might be applied.

Relevance for research

We need to begin by thinking of uncertainty in terms of probabilities of events occurring or of statements being true or false. When evidence or data is available in the form of objective probabilities, then their combination is non-controversial and follows the conventional axioms of probability. However, Bayesians working in the philosophy of science make two controversial assumptions.

The first is the re-definition of probability away from the 'frequentist' view to the subjective view. In other words, instead of a probability being the frequency of occurrence of an event, it is redefined in terms of different degrees of belief in the likelihood of an event or the truth of a proposition. If p=0, it means we believe there is no likelihood of the event occurring, and if p=1, it means we consider the event will certainly happen. Intermediate p values represent the degree of belief (or uncertainty) we have in the likelihood of the event happening or the truth of the proposition. These p values are called 'subjective probabilities'.

The second assumption is that, for a rational person, these subjective probabilities should conform to the axioms of probability theory and can replace conventional probability values in the basic equations on conditional probability.

This broader definition can give us insight into how a new research finding or piece of evidence (E) can alter our credence or acceptance of a theory or research hypothesis (H) associated

QUALITY OF LIFE AND EVIDENCE-BASED BELIEF

with E. For example, if we now look at scientific evidence and its role in hypothesis testing, there are two relationships we might consider:

a) the probability or likelihood of evidence E, given H is true (written as p(E/H));
b) the probability or likelihood of a hypothesis H being true, given evidence E (written as p(H/E)).

The key point is that the first of these probabilities, p(E/H), actually represents the conventional scientific approach to data gathering and hypothesis testing, while the second term, p(H/E), represents a Bayesian approach. In the conventional approach, when researchers gather data (whether from soil measurements or questionnaires), they are expected to subject their findings to conventional statistical analysis. Underlying this evaluation of the evidence or data there is a logic of the form:

a) first set up a null hypothesis (Ho);
b) then assess whether the data you have (D) fits (Ho) . . . p(D/Ho);
c) finally retain or reject (Ho), depending on the likelihood of the data given Ho.

In logical terms, we are assessing the probability of the data being as it is given the hypothesis (Ho). If the value for p(D/Ho) is very low it means the data is unlikely to fit the hypothesis and we reject (Ho) and recommend an alternative hypothesis.

On the other hand, the key insight in Bayesian logic is to invert this expression and to ask, 'What is the probability of the hypothesis being true, given the data?' In other words we use p(Ho/D). By flipping the probability, the Bayesian approach concentrates immediately on the alternative hypotheses of interest, and of course there can be many of these in any research project. So, for example, we might have for any set of data D1

P(Ho/D1) and in addition
p(H1/D1)
p(H2/D1)
p(H3/D1) etc.

Each of these hypotheses receives greater or lesser support from the evidence (i.e. the available data D1) so the key research strategy becomes an assessment of which hypotheses receive greatest support. In addition, this highlights a key characteristic of useful evidence, which is discrimination not association. In other words, a piece of evidence can be strongly associated with two alternative hypotheses but of little value in helping us discriminate between them. For example, in medicine, the presence of a high temperature may be associated or correlated with more than one disease and is on its own no use, therefore, in deciding between them. The key is to find discriminatory evidence. So if a design tutor interested in educational research discovers a strong link between an initial drawing task (T) and better design outcomes (D+) in a student, we need to know whether or not doing the drawing task T (or nonT experience) can lead to (D+) before we can draw the conclusion that there is something special about the role of drawing leading to better designs.

Of course, in any real research project, the hypothesis of interest is linked to many pieces of evidence, D1, D2, D3 and so on. There is no problem in extending the basic ideas or the mathematics to cover this (i.e. we write p(H1/D1,D2,D3 . . .)) although the main focus here is on the Bayesian idea.

Another well known aspect of Bayesian thinking is what is called 'the use of priors'. Bayesian research assumes that any research finding should be assessed not in isolation but on the basis of what has gone before (i.e. the 'priors'). The priors are controversial because, in a Bayesian analysis, you are asked to estimate what the collective weight of research opinion is for H1, H2, H3 and so on, before you carry out your research. Then, by applying the Bayesian formula, the new research data is modified by these priors to show the contribution of the new evidence to the revised credibility of H1, H2 and H3. The greatest p value associated with these alternatives represents the best believable hypothesis at that stage. This is an evolutionary view of research progress in which the priors are shifting over time as new evidence accumulates. So, for prior belief Hp, the evidence E can either

a) be neutral to the current view Hp (i.e. Hp remains the same with or without E) – we might say the evidence is irrelevant to the degree of belief in Hp;

b) reinforce the current view Hp (i.e. Hp is increased in probability and therefore degree of belief); or
c) go against the current view Hp (i.e. Hp is reduced in probability – our degree of belief is less strong).

The advantage of this approach is that research findings are automatically in context. What is considered research could either be findings which reinforce the status quo when this belief is tentatively held, or it could be a shift of belief between possible alternative opinions. This is in contrast to problems for a naïve Popperian approach in science – scientists famously did not ditch hypotheses on the grounds of single experiments which falsified them but sensibly waited for evidence to accumulate from several sources (Harre, 1986). Such scientific practice is in line with a Bayesian approach. Any research project is carried out against a background of similar research. This is, of course, reflected in the literature survey researchers are obliged to write, but it is *not* reflected in the conventional analytical procedures.

Landscape applications

Let us suppose I am interested in Ulrich's (1984) claims about the beneficial effects of viewing landscape scenes in hospital. These findings have been published but problems have been encountered over their replication, and we might suppose that scientists are divided over Ulrich's and colleagues' claims. Let us consider the problem from a Bayesian perspective.

From the Bayesian formula

$$p(H/E) = p(H) \times p(E/H)/\{ p(H) p(E/H) + p(-H) p(E/-H)\}$$

where p(H/E) probability of H being true given E
p(E/H) probability of E occurring given H is true
p(E/-H) probability of E occurring given H is not true
p(H) prior probability of H being true
p(-H) prior probability of H being not true

a) Bayesian significance without conventional significance

Initially I estimate the priors. If I am genuinely agnostic or uncertain about the claim (and let us suppose the literature

Table 13.5 A simple Bayesian estimation

Research hypothesis, question or belief	Priors	Evidence	
p(H1/E1)	0.5	Probability of Evidence E1 given H1	0.9
p(-H1/E1)	0.5	Probability of Evidence E1 given -H1	0.3

doesn't support one view over another), I can set the two priors (p(H) and p(-H)) to 0.5 for the two-alternative case (i.e. I don't know) and carry out my research study. The best way to see how the probabilities are combined is to set them out in a table (see Table 13.5).

The left column represents my main research question

a) p(H1/E1) – what is the probability of H1 being true given evidence E1 (i.e. the probability that landscape scenes do reduce stress, given the evidence in the Ulrich publication);
b) p(-H1/E1) – what is the probability of H1 being false given evidence E1 (i.e. the probability that landscape scenes do not reduce stress, given the evidence in the Ulrich and Simons publication).

The second column gives my priors set at 0.5. The fourth column reflects the likelihood of the evidence E1 given H1 was true, and below it if H1 was false. Let us assume these values are 0.9 and 0.3. This assumption might be based on the subjective judgement of one person, or on the collective judgement of several professionals, or on the outcome of a number of experimental studies.

From Bayes

$$p(H/E) = 0.9 \times 0.5/ (0.9 \times 0.5 + 0.5 \times 0.3) = 0.45/ (0.45 + 0.15) = 0.75,$$ and it follows that $p(-H/E) = 0.25$

In other words, given my agnosticism and the evidence, even if data has not reached conventional significance (i.e. we assumed a value of 0.9 with an error rate of 0.1 rather than a conventional p value of 0.05), we accept the claim of Ulrich against the alternative hypothesis credibility of 0.25. Furthermore, if someone else carries out research in this area the new priors become 0.75 and 0.25. In other words we are moving away from total agnosticism of 0.5 and 0.5 towards moderate support for the Ulrich claim.

b) Non significance for conventional and Bayesian analysis, with the same data and shift of priors

Suppose I am more sceptical about the claim before the Ulrich study and think there is only a 20% chance of it being true (i.e. I set the priors at 0.2 and 0.8). Then

$$p(H/E) = 0.9 \times 0.2/ (0.9 \times 0.2 + 0.5 \times 0.8) = 0.18/ (0.18 + 0.40) = 0.31 \text{ with } p(-H/E) = 0.69$$

Now, under the same evidential data, we do not accept Ulrich's claim and instead accept the alternative until, once again, new evidence emerges. If the generation of priors appears too uncertain in the brief above account, it is important to note that:

a) there are decisional techniques for the generation of subjective probabilities (Tversky and Kahneman, 1982);
b) there is a convergence towards more objective data after a few empirical studies have been undertaken.

The Bayesian approach can easily be integrated into decisional frameworks. For example, using the Delphi technique (Kirk and Spreckelmeyer, 1988), initial disagreement could be reduced so that a professional consensus view could be achieved. This could result in some form of knowledge base for the discipline for those issues which were more controversial. The good news is that convergence occurs fairly quickly, irrespective of initial prior values.

It should be emphasised that this is not a descriptive account of how people reason. There is a wealth of psychological evidence to show that people are not naïve Bayesians and that inferences from probabilistic evidence are error prone (Cave, 2002). What is presented here is a normative view of how inferences might be made. Bayes offers a structure for combining evidence and beliefs which is more flexible and less prescriptive than conventional criteria. It also helps ground these beliefs with sensitivity gauged by the selection of different criteria for inclusion in the formula. A more sophisticated version involving multiple integration underlies the map in Figure 13.5. However, the purpose of this chapter is to suggest Bayesian ideas as a philosophical framework for research in complex areas of uncertainty which depend on skilled intuitive judgements. In these circumstances a Bayesian approach is a way of building an inferential knowledge base on a mix of evidence and subjective beliefs. Bayesian ideas may be a constructive way forward for interpreting research findings and for research development.

Conclusion

An outline has been given of three different OPENspace projects in which more recent quantitative methodologies have been used. In the first part of this chapter, an example on quality of life was presented using conjoint analysis to address the issues of personal difference in priority and difficulty for people with visual impairment. It will be evident that this approach is of potential value to a wider range of research projects associated with the evaluation or preference for different landscapes and environments. One of the main points arising is that landscape research on preference would benefit from more rigorous methodologies involving relative choice. Following this, pictorial regression was illustrated using AnswerTree. This analysis offers the researcher exploratory regression with the researcher in control and the process transparent, rather than a 'black box' procedure. Different variables can be introduced into the tree structure and their impact assessed on the predictive value of the tree. In completing the illustration of quantitative methods, latent class analysis was applied to reveal evidence of two different types of visitor to woodlands. Finally, an alternative Bayesian way of relating evidence to belief is presented in the hope that it provides insights into research findings and their credibility.

References

AnswerTree – module supplied by SPSS Inc (see www.spss.com).

Aspinall, P., Hill, A.R. and Nelson, P. (2005) 'Quality of life in patients with glaucoma: a conjoint analysis approach', in *Journal of Visual Impairment Research*, vol. 7 (1), p.13.

Aspinall, P., Hill, A., Dhillon, B., Armbrecht, A.M., Nelson, P., Lumsden, C. and Buchholz, P. (2007) 'Quality of life and relative importance: a comparison of time trade-off and conjoint analysis methods in patients with age-related macular degeneration', in *British Journal of Ophthalmology*, vol. 91, pp. 766–772.

Cave, S. (2002) *Applying Psychology to the Environment*. London: Hodder and Stoughton.

Freedman, L. (1996) 'Bayesian statistical methods', in *British Medical Journal*, editorial, 7 September, vol. 313 (7057), pp. 603–607.

Harre, Rom (1986) 'Persons and powers', in S.G. Shanker (ed.) *Philosophy in Britain Today*. New York: State University of New York Press, pp. 135–153.

Hill, A. and Aspinall, P. (2004) 'Satisfaction as an outcome measure of low vision services', in *Journal of Visual Impairment Research*, vol. 5 (3), p. 1.

Hyde, M., Blane, D., Higgs, P. and Wiggins, R. (2003) 'A measure of quality of life in early old age: the theory, development and properties of a needs satisfaction model (CASP-19)', in *Aging & Mental Health*, vol. 7 (3), pp. 186–194.

Johnson, R. (2001) *Conjoint Analysis Software*. Sawtooth Software Inc., 530 West Fir St., Sequim, WA 98382, USA (see www.sawtooth software. com).

Kirk, S.J. and Spreckelmeyer, K.F. (1988) *Creative Design Decisions: a systematic approach to problem solving in architecture*. New York: Van Nostrand Reinhold.

Leishman, C., Aspinall, P. and Munro, M. (2004) *Preferences, Quality and Choice in New Build Housing*. York: Joseph Rowntree Foundation.

Little, B.R. (1998) 'Personal project pursuits: dimensions and dynamics of personal meaning', in P.T. Wong and P.S. Fry (eds) *The Human Quest for Meaning*. Mahwah, NJ: Lawrence Erlbaum.

Mangione, C.M., Phillips, R.S., Seddon, J.M., Lawrence, M.G., Cook, E.F., Daily, R. and Goldman, L.G. (1992) 'Development of the activities of daily vision scale. A measure of visual functional status', in *Medical Care*, vol. 30, pp. 1111–1126.

Mangione C.M., Lee, P.P., Gutierrez, P.R., Spritzer, K., Berry, S. and Hays, R.D. (2001) 'Development of the 25-item National Eye Institute Visual Function Questionnaire', in *Archive of Ophthalmology*, vol. 119 (7), pp. 1050–1058.

Massof, R.W. (1995) 'A systems model for low vision rehabilitation. I. Basic concepts', in *Optometry and Vision Science*, vol. 72, pp. 725–736.

National Institute of Clinical Excellence (NICE) (2004). *Guide to the methods of technology appraisal*. Report, April (Section 5.5: Valuing health effects, p. 24). London: Department of Health.

Orme, B. (1998) *Sample Size Issues for Conjoint Analysis Studies*. Sawtooth Software. Research paper series. Sawtooth Software, Inc. 530 West Fir St., Sequim, WA 98382, USA.

Rubin, G.S., Muñoz, B., Bandeen-Roche, K. and West, S.K. (2000) 'Monocular versus binocular visual acuity as measures of vision impairment and predictors of visual disability', in *Investigative Ophthalmology and Visual Science*, vol. 41 (11), pp. 3327–3334.

Rummlab Pty Inc, see www.rummlab.com.au.

Statistical Innovations, Latent Class, see www.statisticalinnovations.com.

Tversky, A. and Kahneman, D. (1982) 'A heuristic for judging frequency and probability', in D. Kahneman, P. Slovic and A. Tversky (eds) *Judgement Under Uncertainty: heuristics and biases*. Cambridge: Cambridge University Press, pp. 163–178.

Ulrich, R.S. (1984) 'View through a window may influence recovery from surgery', in *Science*, vol. 224, pp. 420–421.

Ward Thompson, C., Aspinall, P., Bell, S., Findlay, C., Wherrett, J. and Travlou, P. (2004) *Open Space and Social Inclusion: Local Woodland Use in Central Scotland*. Edinburgh: Forestry Commission.

Ward Thompson, C., Aspinall, P., Bell, S. and Findlay, C. (2005) '"It gets you away from everyday life": local woodlands and community use – what makes a difference?', in *Landscape Research*, vol. 30 (1), pp. 109–146.

Index

Page numbers in *italics* refer to figures.

Abbeyfield Park, Sheffield *47*
access: and participation, ethnic minorities 41, 48–50; *see also* disabilities; wayfinding
affordance, concept 87–8, 101, 127–9
ageing 153–4
Alzheimer's disease 137–50: amygdala 138; assisted living 138, 142, 146; characteristic behaviours 149; design schema for people with 143; frontal lobe 138; hippocampus 137; Multidimensional Observation Scale for Elderly Subjects 149; Reisberg Psychotic Symptom List 149; suprachiasmatic nucleus (SCN) 138; *see also* healing gardens
AnswerTree analytical technique 185–6, *187*; classification and regression tree (CART) 186
Anti-Social Behaviour Orders (ASBOs), UK 34, 72
antidepressants (SSRIs) 175, 176
Appleton, J. xiv
attention restoration theory 165–9, 170–1

Bankside Open Spaces Trust, UK 50, 52
Bayesian: analysis 181, 184; approach xix; inference 189–93
behaviour: Alzheimer's disease 149; Anti-Social Behaviour Orders (ASBOs) 32, 72; mapping 88, 89, *90*, *91*; settings 87, 88, *90*, *91*, 95, 96, 97, 98, 99, 105 and setting type user profiles 99–100; tracking 88, *89*
Bell, S. *et al.* xviii, 25, 29, 30, 71
Bestwood Country Park, Nottingham, UK 50, *51*
Black Environment Network (BEN) 43–5, 46, 50, 52
Bristo Square, Edinburgh *see* Edinburgh
British attitudes to teenagers 71–2
British Heart Foundation 18, 23
British Trust for Conservation Volunteers (BTCV) 20, 43–5
Britishness and multiculturalism 41–2, 45–6
Brockwell Park Lido, London 13

built environment: and active living 126–7; public health movement 13–14
Busmanis, P. *et al* 67, 68

CABE *see* Commission for Architecture and the Built Environment
Calthorpe Project, London 50, 53
Canada 85
cancer 172
Canter, D. 34, 57, 66, 73
Carson, R. xiii
Cary Parks Commission, US 86
Cashel Forest, Scotland *44*
Central and Eastern European (CEE) countries 56–7; Latvia 57, 61–6, 67–8
Chartered Institution of Water and Environmental Management (CIWEM) 43
Chumleigh Gardens, Burgess Park, London 47–8, *48*
child/adult ratio (CAR), park user profiles 99–100
childcare centres, outdoor play environments 130–1
childhood connections: with outdoor places 26–9, 36; with rural Latvia 63, 65
childhood obesity 125–6, 132–3
children: developmental needs 24, 86, 88, 125–33; with disabilities 86, 101–4, 105, 108; Kids Together Parks study, US 86–108; from low-income housing 171; universal design 86
Cimprich, B. and Ronis, D.L. 172, 177
citizenship 45, 46
city space: café culture in 6, 7; CIAM Athens charter 4; early 1960s 4–5; function of as connection space, marketplace, meeting place 3–4, 8–9; impact of the car 4, 6; impact of technology 6, 7, 8; increase in city life, Melbourne, Australia 6; Modern Movement 4; pedestrianisation 4–6, 8; public life in 3–4; quality of 3–6; in the twenty-first century 8–9; *see also* Copenhagen; Edinburgh; London; Riga
Clackmannanshire, Scotland, visitor sign *27*
Clarion Cycling Club, UK 14

Classification and Regression Tree (CART) *see* AnswerTree analytical technique
cognitive mapping 138, 143–4
Cohen-Mansfield Agitation Inventory 149, 150
Commission for Architecture and the Built Environment (CABE) 18, 20, 23, 36
Commission for Racial Equality 47, 52
Commission for Rural Communities 56, 68
commodity, notion of xiii
conjoint analysis, choice-based 181, 182, 185
Conn, D. 19
Consumer Product Safety Commission (CPSC), US: guidelines 132
contested spaces, teenagers in Edinburgh 74–80
cool summer weather 174–6
Cooper Marcus, C. xiv, 87, 109, 169, 177
Copenhagen 4, 5–8
Cosco, N. xviii; Cosco, N. and Moore, R. *Childhood Outdoors . . . North Carolina Childcare Centers* 130
Countryside Agency 23, 42, 52
Cullen, G., 14
cultural activities: ethnic minorities 47–8, 50, 52, 105; Scotland 59
'cultural gardens' 47–8, 50
cultural and social inclusion 41–52, 85–6, 105–6, 107, 108
culture, youth 72, 73, 75
cumulative effects of repeated restorative experiences 169–73
cycling *17*, 155

danger *see* fear; safety
Department for Communities and Local Government, UK 19, 23
Department for Environment, Food and Rural Affairs (DEFRA), UK 42
depression 175–6
design: accessible 85–108; ethnic minorities access and participation 50; evidence-based, child development in outdoors 125–33; for inclusive access xvii–viii; Inclusive Design for Getting Outdoors (I'DGO) 153–62; Kids Together

INDEX

Park study, US 86–108; universal 85–6, 108
dialogue, ethnic minorities access and participation 49
disabilities 108; children with 86, 101–4, 105, 108
Disability Discrimination Act, UK 86
discrete restorative experiences 165–9
Downing, A.J. 165
Downs, R.M and Stea, D. 113, 123
Drache, D. 85, 108
Durham Heritage Coast, UK 111

East Midlands, UK 25–6, 29
Edinburgh: intergenerational usage of open space 34; playing outdoors 28; skateboarding in Bristo Square 77; visual impairment study 181–6, *187*; youth study 72–80
employment, rural Latvia 64, 67
English Nature 25, 30
entrance reassurance, 'Site Finder' toolkit 120–1
environment–physical activity interaction 154–6
Environmental Design Research Association (EDRA) xiii
environmental attributes (SEA) 156–7, 158, 160–1
environmental supportiveness measures 156–8, 160–1
environmental survey, wayfinding study 115–16
Esplanade, White Plains, NY *see* Alzheimer's disease, *and* healing gardens
Europe xii: poor design of environments in xii; rebuilding in 1920s–30s xiii; rural social exclusion 55–68; urban development in, postwar to 1990s xiii–xv
European Union (EU) 55, 67; action on diet, physical activity and health 126
evidence-based approaches 125–33, 181–93
exercise *see* physical activity; play

family visits to parks 89–91, 104–5
fear: of bullying 78–9; of crime 35, 47
Finsbury Health Centre, London, 13
fitness and exercise 11–12, 14, 16, *17*, 18–19, 30
Forest Authority 112
Forestry Commission 24, 111, 112, 123
forests *see* woodlands
'form follows function' xiii
freedom and control, use of parks 107
functional planning xiv
functional zones, parks 88, *90*, 91–4, 107; attractiveness index 94–8; children with disabilities 101–4
funding policy, ethnic minorities access and participation 49–50

gardens *see* 'cultural gardens'; healing gardens
Gartmorn, Scotland *27*
gatekeeping, healing gardens 148
Geddes, P. xiii
gender, park user profiles 100
Gehl, J. xiv, xviii
Geographical Information System (GIS) 89, 91, 155
Germany 14
Gibson, E. and Pick, A. 88, 127, 129
Gibson, J. 87, 88
Giles-Corti, B. *et al* 164. 178
glaucoma 182, 184; visual impairment study 181–6, *187*
'goths' 75, 76
Grant, C. 147, 148
green spaces, UK government policy 12, 19–20, 23
GreenSpace (Urban Parks Forum), UK 15
Groundwork, UK 20

Halprin, L. xiii
Hannaford, C. 86, 125, 126
Hartig, T. xviii; *et al.* 164, 167, 169, 173, 175
healing gardens 138–46; Alzheimer's disease 137–8, 146, 149; evaluation of effects 146–50
health issues: childhood obesity 125–6, 132–3; ethnic minorities 47; public health, UK 11–20, 23; *see also* restorative environments
health: mental and physical 163–177; *see also* restorative environments; healing gardens
healthcare settings, restorative experiences 171–2
Hearthstone Alzheimer Care: Mass, US 138–42; NY, US 142–6
Heft, H. 35, 87, 88, 128
Heritage Lottery Fund 18, 45
Hester, R. xiv
hidden groups analysis, use of local woodland 187–9
Hillier, B. xiv
historic development of cities 3–9
House of Commons, UK: ASBOs 72; Select Committee 'Inquiry Into Obesity' 12, 18
housing: low-income, children from 171; multi-family public 171
housing zones, universal 144–6, 147, 148
Hunt, T. 11

identity *see* sense of identity
inclusion, social and cultural 41–52, 85–6, 105–6, 107, 108
Inclusive Design for Getting Outdoors (I'DGO) 153–62
indoor leisure and fitness activities 16, 18
informational consistency, 'Site Finder' toolkit 119–20
instrumental activities of daily living (IADLs) 157
intergenerational issues 34, 64–5, 67
interview survey, wayfinding study 114

Jackson, J.B. xii, xvi
Jacobs, J. xiii, 4, 9, 145, 150
Jakle, J.A. 113, 124
Jellicoe, G. 14
Jones, G. and Greatorex, P. 16, 18

Kaplan, R. 171; and Kaplan, S. 165–6
Kaplan, S. 166
Kelly, G.: personal construct theory 57, 155
Kids Together Park study, US 86–108
Kuo, F.E. and Sullivan, W.C. 170, 171

INDEX

Land Use Consultants 23, 37
landmarks 113, 144
latent class analysis 185, 187; use of local woodland 187–9
Latvia 55, 57, 61–6, 67–8
leisure and fitness activities 16, *17*, 18–19, 30
Lieberg, M. 71, 80
Lister Park, Bradford, UK *see* Mughal Garden, Bradford
Little, B. R. 35, 154–5, 156, 162, 181, 193
'Lonach', Scotland 59, 66
London 13, 14, 18–19, *20*, 47–8
Lubetkin, B., 14
Lynch, K. 113, 143–4, 147

Mace, R. 85–6
McHarg, I. xiii
macular degeneration 182, 184; visual impairment study 181–6, *187*
Madgely, J. and Bradshaw, R. 56
Manchester Commonwealth Games 2002 19
mapping: behaviour 88, 89, *90*, 91, 100, 110; cognitive 138, 143–4; place, and teenage groups 73–4, 79–80
Mason, G. *et al* and behaviour mapping 100, 110
Maryon Park, London *15*
Matless, D. 11, 15, 21
Matthews, H.: *et al.* 71–2, 78, 80; Limb, M. and 74, 80; Percy-Smith, B. and 78, 79, 80
microgeographies: teenagers' creation of 73, 78
Mile End Park, London *20*
modernist movement 4, 13–14
Moore, R. xviii, 24, 87, 100; *et al.* 86, 132–3; and Wong, H. 100, 125, 126; and Young, D. 87
Morris, W. 14
Mugdock Country Park, Scotland 44
Mughal Garden, Bradford 50, *51*
Muir, J. 165, 179
multicultural interpretation 45–8, 50
multiculturalism and Britishness 41–2, 45–6

National Council for Voluntary Organisations, UK 46, 52
National Health Service, UK 18, 19
National Institute for Clinical Excellence (NICE), UK 181
National Institute of Health (NIH), US 126
National Parks Bill, 1949, UK 15
National Playing Fields Association, UK 16
National Survey of America's Families, 1999 126
National Trust, UK 45
natural environments 24–6; childhood 24, 26–9; intergenerational issues 34; older people 31–3, 34; teenagers 30–1, 34, 36; *see also* woodlands
Neighborhood Environmental Walkability Scale (NEWS) 155, 156
New Labour 16, 18
Newman, O. xiii
Nikodemus, O. *et al* 65, 68
Norman, D.A. 138, 143, 147
North Carolina, US: childcare centers 130–131
North Willen Park, Milton Keynes, UK 46

obesity 12; childhood 125–6, 132–3
observations 116; setting 88, 89
Office of the Deputy Prime Minister (ODPM), UK 12, 19
older people xv, 23, 31–33, 34, 35, 46, 137–150, 153–161; and Alzheimer's disease *see* healing gardens; challenges for 154; in cities 35; Inclusive Design for Getting Outdoors (I'DGO) 153–62; intergenerational issues 34, 64–5, 67; natural environments 31–3, 34
Olmsted, F.L. xii, xiii, 165, 173, 179
Olympics, London 18–19
open space/outdoor environments: childhood memories of 26–29; community health xv; current theory and practice in, xvii; democracy and social equity in relation to, xvii, 41–52; design challenges of xvii, 23, 85–108; evidence-based design 125–133, 137–150; evidence-based research 181–193; healing gardens 137–150; and health 11–21, 125–133, 137–150, 163–177; immigrant communities, perceptions of xv, 41–52; inclusive access to xvii, 41–52, 55–68, 71–80, 85–109; intergenerational issues 23, 33, 34, 64–5, 67, 71–80; organization of xv; xvii; physical attributes of xv; people's engagement with 23–36, 71–80; public life in cities 3–9, 11–20, 71–80; public policy, with regard to 11–20; quality of life in relation to xvii; 55–68, 153–161, 181–193; restorative environments 163–177; rural social exclusion (Europe) 55–68; universal design 85–109; value of play in 23–24, 85–109; wayfinding in 111–123; wellbeing xv; xvii, 137–150, 163–177; woodlands 24, 25, 26, 29, 30, 31, 32, 34, 35; *see also* parks
open space, users' engagement with: children 23–24, 26, 32, 33, 36, 71, 85–109, 125–133; children with disabilities 101–105; city dwellers 3–9; minority ethnic groups 25–26, 41–52; older people xv, 23, 31–33, 34, 35, 46, 137–150, 153–161; over-55s 25; people (general) 11–20; 23–36, 85–109; people with Alzheimer's 137–150; people with health problems, 163–177; rural dwellers, 55–68; teenagers xv, 23, 24, 30, 31, 32, 34, 35, 36, 64, 71–80, 106; visually-impaired people 181–86; visitors to the countryside 111–123; women 25, 46; young people 25, 34, 56, 64, 71–80
OPENspace Research Centre, Edinburgh College of Art 24, 25, 26, 30, 31, 72, 181, 185, 186, 188, 193
organizational development, ethnic minorities access and participation 49
Ostroff, E. 85; and Iacofano, D. 85–6

parks 12–18, 19–20, 165; Kids Together Parks study, US 86–108; safety in 106, 107; Victorians' provision of 11
partnership, ethnic minorities access and participation 49
Passini, R. 112, 113
Peak District National Park, UK 111
Peckham Health Centre, London 13
pedestrian environment 4–6, *8*

INDEX

'people, purpose, place' model of signage 116–17
Percy-Smith, B. and Matthews, H. 78, 79
Perrin, C. xiii
personal construct theory 35, 57, 155
Personal Projects, concept of 35, 154–5, 156, 158–60, 161
physical activity: and environment interaction 153–6; and fitness 11–12, 14, 16, *17*, 18–9, 30; planned programmes, healing gardens 148
Pikora, T.J. *et al*: Systematic Pedestrian and Cycling Environmental Scan (SPACES) 155
place-mapping and teenage groups 73–4, 79–80
play 129–30; outdoors, childcare centres 130–1; and safety regulations 131–2
population decline, rural areas 56–8, 61–2, 65, 67
Preschool Outdoor Environments Measurement Scale (POEMS), US xviii, 133, 134
Prestopnik, J.L. and Roskos-Ewoldsen, B. 112
probability (Bayesian inference) 189–93
problem analysis toolkit 112
professional development, ethnic minorities access and participation 49
psycho-evolutionary theory 166–7, 170–1; stress and blood pressure reduction 166, 167, 168–9
public health, UK 11–20, 23
Public Health Service Department, Southwark, London, 13
public life: erosion of 4–5; new roles 8
'Public Park Assessment', UK 15
public parks *see* parks

Qingdao, China, playing outdoors *29*
quality of life: for older people 154–61; and social exclusion in rural areas 55–69; for visually impaired people 181–6, *187*

racism 46, 47
Ramblers' Association 15
Rasch analysis 185

regression analysis (AnswerTree) technique 185–6, *187*
Research Triangle Park, NC, US *127, 128, 129, 130 see also* children, developmental needs 125–133
restorative environments 163–4, 177; cumulative effects 169–73; discrete experiences 165–9; health resource 163–177; social ecological influences 173–6; *see also* healing gardens
Riga 62, 67–8
role-play and spatial analysis 115
Roots Culturefest, Northamptonshire, UK 47
route connectivity, 'Site Finder' toolkit 120
Rowbothan, H.A., 13
rumination 175–6
rural landscapes: Latvia 57, 61–6, 67–8; Strathdon, Scotland 57–60, *61*, 65–6, 67–8
Rural Poverty and Exclusion Working Group 56

Saelens, B.E. *et al*: Neighborhood Environmental Walkability Scale (NEWS) 155
safety: and danger, teenagers' perceptions 74–80; family park visits 106; regulations and play value 131–2
Sallis, J. *et al*. 125, 129, 133
Sawtooth software 182–3, 184
Scharf, T. and Bartlett, B. 56
Scotland *28, 31, 44*, 55, 56; 'Lonach' 59, 66; sedentary lifestyles in childhood 125; Strathdon rural area study 57–60, *61*, 65–6, 67–8; woodlands 24–5, *26, 27, 29*, 30, 31–2; *see also* Edinburgh
Scott, M.J. and Canter, D.V. 66
Scottish Executive 56, 68
Scottish Natural Heritage 23
Sebba, R. and Churchman, A. 78
sedentary lifestyle 125–6
sense of belonging 45, 48
sense of identity: Latvia 63, 68; Scotland 59, 60, 68; youth 71, 75, 79
setting observations 88, 89, 96
setting types 98–101
Siegl, A. and White, S. 113, 124

signs *see* wayfinding
Silkin, L. 14–15
site legibility, 'Site Finder' toolkit 118–23
skateboarders 75–8
sleep 164
sleep-wake cycle, Alzheimer's disease 138, 146
Smithson, T.L., 13
social cognitive theory 34–5
social cohesion policies 42–3
social and cultural inclusion 41–52, 85–6, 105–6, 107, 108, 112
social development, responsibility for 42–3
social ecological influences on restoration 173–6
social exclusion 16–18, 46–7, 56–7, 63
social–physical environment interactions 66, 67
socio-economic context, rural areas 56–7, 63
spatial analysis and role-play 115
Sports Lottery Fund 19
Statistical Innovations software 185, 187
Stockholm, parks *17*
Strathdon, Scotland *see* Scotland
stress: and blood pressure reduction 166, 167, 168–9; and depression 175–6
Sugiyama, T. and Ward Thompson, C. 23, 33, 35
Sweden *17*, 174–6: Swedish National Energy Administration (STEM) 174, 176
Systematic Pedestrian and Cycling Environment Scan (SPACES) 155

Taylor, A.F. *et al* 171, 179
Taylor, H. 14
teenagers xv, 23, 24, 30–1, *32* 34, 35, 36, 64, 71–80, 106; Edinburgh study 72–80; intergenerational issues 34, 64–5, 67; use of parks 106; use of public spaces 71–2
teleworkers, social ecological influences on restoration 173–4
territorial range development concept 87, 88
theory of navigation 113 *see also* wayfinding
Theory of Place 57
tourism, Scotland 59, 60
travel diaries analysis, visually impaired people 185–6, *187*

198

INDEX

Travlou, P. xviii, 23, 30, 36, 71, 72, 73; *et al.* 73
Tuan, Yi-Fu xiv
Tyson, M.M. 147

Ulrich, R.S. 166, 171–2, 179, 191–3; *et al.* 167
United States (US): childhood obesity 125, 126; Consumer Product Safety Commission (CPSC) guidelines 132; Kids Together Parks study 86–108; National Endowment for the Arts (NEA) 85–6; *see also* healing gardens
universal design 85–6, 108
urban environments *see* built environment; cities; Copenhagen; Edinburgh; London; Riga
Urban Green Spaces Taskforce, UK 16–18
US Surgeon-General's Report on Physical Activity 126, 135
use/space ratio (USR), parks 100–1
user profiles, parks 99–100

vacation legislation, Sweden 175
vegetation 126–9, *130*, 148
Vemuri, A. and Costanza, R. 66, 68
visual impairment study 181–6, *187*
Vitruvius xii, xiv, xvi: *Ten Books on Architecture* xii, xvi
Volicer, L. 149, 150
Vondel Park, Amsterdam *17*

'walkability' 156–7
walking 155
Ward Thompson, C. xviii, 111–12; *et al.* 23, 24, 29, 30, 36, 71, 112, 117; and Scott Myers, M. 59
wayfinding: fundamentals 112–13; study 114–23; systems 112; visitor information needs 113–14
Williams, Sir Owen, 13
Wong, J.L. xviii, 45, 46, 50

woodlands 34; Khalsa Wood 50, *51*; Scottish 24–5, 26, *27*, 29, 30, 31–2, 59; use of, latent class analysis 187–9; *see also* wayfinding
workplace, restorative experiences 171, 173–4
Worpole, K. 11, 13, 14, 18, 36

youth: culture 72, 73, 75; identity 71, 75, 79; *see also* teenagers

Zeisel, J. 149; *et al.* 146; and Tyson, M. 143; and Welch, P. 145, 147, 148
zones *see* functional zones, parks; housing zones, universal

An environmentally friendly book printed and bound in England by www.printondemand-worldwide.com

PEFC Certified
This product is from sustainably managed forests and controlled sources
www.pefc.org
PEFC/16-33-415

Mixed Sources
Product group from well-managed forests, and other controlled sources
www.fsc.org Cert no. TT-COC-002641
© 1996 Forest Stewardship Council

This book is made entirely of chain-of-custody materials

#0460 - 151111 - C0 - 276/216/12 - PB